SIDEWINDER

SIDEWINDER

Creative
Missile
Development
at
China Lake

ron westrum

Naval Institute Press / Annapolis, Maryland

Naval Institute Press
291 Wood Road
Annapolis, MD 21402

Library of Congress Cataloging-in-Publication Data
Westrum, Ron, 1945–
    Sidewinder : creative missile development at China Lake / Ron
    Westrum.
        p.   cm.
    Includes index.
    ISBN 1-55750-951-4 (alk. paper)
    1. Sidewinder (Missile)—Design and construction. 2. Naval Weapons
Center. I. Title.
UG1312.A6W47 1999
623.4'5191'0973—dc21   99-27338

Printed in the United States of America on acid-free paper ∞
06 05 04 03 02 01 00 99   9 8 7 6 5 4 3 2
First printing

To the men and women of China Lake

Contents

Illustrations

Preface

Why China Lake?

This book began innocently enough in 1986 when one of my students, Steve Wilcox, noticed a *Wall Street Journal* article about a guided missile. I recall Steve urging me to read the article, in spite of my protests that I didn't care about guided missiles. When I read it, I found the story irresistible. In 1956 a small team of men in a remote government laboratory created a guided missile that no one had asked for, a guided missile that soon became a world standard. Underdog stories have a fascination for me, and I began to research "the little missile that could." Before long I realized the real story was not the missile but the people who created it. China Lake was a dynamic environment. Sidewinder, its best-known accomplishment, was only one among many. Little did I realize then that this book would be interwoven with the strands of my life for thirteen years.

Understanding China Lake became my goal. While the Sidewinder missile is the focus of this book, my intent is to show how this highly creative environment worked. Started in World War II as the Naval Ordnance Test Station (NOTS) (later the Naval Weapons Center [NWC]), China Lake developed weapons. It specialized in systems used by aircraft, but its contributions reached into many areas.

Its design and structure had one purpose: to foster technical creativity. It did; China Lake operated far outside the normal envelope. Sidewinder was the high point, but there were many other "impossible" accomplishments that China Lake carried off with aplomb. Its ability to do such things inspired all those who shared the common enterprise and sweat blood to make it function. Phrases like "the golden days," "Camelot," "the best time in my life" came easily to the lips of those who related these events. And what was golden about it? The ability to make ideas triumph.

I remember the interview with Hack Wilson. It didn't start well. H. G. Wilson had served as associate technical director of China Lake under Bill McLean and had acted as technical director himself. Retired now, he met me at the door of his home and said, "I told Howie I didn't think this was a good idea." But he invited me in, and we spoke for about two hours about China Lake and Bill McLean and what it all meant. As I left he gripped my arm. Hack was glowing. "You're doing something very important, young man," he said, "Go to it!" He had remembered China Lake.

Bill McLean and his ideas are woven through the narratives of this book. McLean not only developed the Sidewinder, he was China Lake's most famous technical director, serving from 1954 to 1968. McLean did not create China Lake, but he helped shape it for more than two decades. When he left, China Lake would not be the same; neither would Bill McLean. China Lake suited Bill McLean perfectly. Although he did not run out of ideas when he left and moved to San Diego, fewer of his visions got realized at the Naval Undersea Center, where he became technical director. China Lake's special qualities proved harder to duplicate than Bill McLean had realized.

Sidewinder was a logical outgrowth of China Lake's approach to R&D. Just as China Lake supported Bill McLean, it also provided a nurturant environment for his concept. Sidewinder was a better missile than its competitors. That superiority was no accident. It reflected the style of R&D practiced by China Lake. By allowing Bill McLean to do what he thought was right rather than giving the navy what it asked for, China Lake created a world-class weapon. And it created many other fine systems in the same proactive, hands-on way. China Lake had technical brilliance backed by courage. Its achievements are testimony to the value of such brilliance and such courage.

This is by no means the whole story of China Lake. That is being told through a fine series of books by historians of NOTS and NWC. The third volume, covering much of the period I discuss, was being written by Elizabeth Babcock as this book went to press. This book concentrates on how China Lake was put together and how it worked. Understanding how it operated can be a guide in creating similar places elsewhere. China Lake's high standards, deep knowledge, visionary leadership, and freedom to inquire allowed it to achieve "utter probity toward the engineered object" to use the phrase of Arthur Squires.[1] I have written this book to describe how they did it.

Sources

This book was written largely from the recollections of the members of the Sidewinder project and other projects associated with it or with Bill McLean. I have also used records obtained either from the participants or from official sources. Most of the interviews were carried out by me; most of the early ones, with the assistance of Howard Wilcox. I am grateful to the many people who invited us into their homes, hunted up documents, wrote out recollections, and answered our many questions. Historians Liz Babcock and Leroy Doig III, at what is now the Naval Air Warfare Center, Weapons Division, at China Lake, California, supplied me with a great variety of important materials and responded to innumerable questions. Liz Babcock especially went to great lengths to dig up materials and interviews, and her team even conducted some joint interviews with me. I await her own book, the third volume in the official China Lake history. Also helpful were David Allison, former historian of Naval Laboratories; Marc Jacobsen, historian at the former Naval Ocean Systems Center in San Diego; and Joe Marchese, then historian of Naval Laboratories. I am grateful to LaV McLean, who was kind enough to allow me access to the materials in McLean's personal files and to McLean's final notebook. McLean's patent attorney Walter Finch also supplied a number of useful documents.

A limitation on this study is that many of the materials on Sidewinder's history are still classified, as are many other documents it would be useful to have. Other key documents were destroyed in a flood at China Lake in 1984.

This book began in 1986 in partnership with Howard A. Wilcox. Howie later removed himself from the project and died before it could be completed. But in many respects, this book has been shaped by his initial participation, ideas, and insights. Several other people later took on the role of associate historians for this study and acted as coinvestigators. Principal among these was the late Robert Blaise, who spent countless hours giving me essential information, insights, and feedback. Bob's death before completion of the book is deeply regretted. Other senior advisers were Tom Amlie, Burrell Hays, Frank Knemeyer, Ed Paul, Chuck Smith, Conway Snyder, Henry Swift, Glenn Tierney, and Norm Woodall. These men wrote me extensive letters and carried out their own personal investigations. They deserve special mention. They hunted down witnesses, provided conduits for key materials, and proved trenchant critics of my writing. Bud Sewell also helped recruit interviewees. In a later phase of the study, Max White, historian of the Point Mugu Pacific Naval Test Center, was kind enough to share his ideas, materials, and insights with me.

A primary debt is owed to David Rugg, who began a similar enterprise in the early 1970s, only to have it prematurely terminated when funds ran out. Rugg's tape-recorded interviews and his notes have been extremely valuable, especially in those areas for which records are no longer accessible. Other studies such as those written by Theodore Gautschi and Thomas A. Marschak were also valuable.

Many individuals were helpful with advice, photographs, records, and clippings. Among these are Stanley Atchison, Ernie Cozzens, Fred Davis, Walt Freitag, Walt LaBerge, Tom Lang, Jack Rabinow, Leroy Riggs, Dick Schmitt, Glenn Tierney, and Gene Younkin. Bob Blaise, Barney Smith, and Howie Wilcox wrote extensive personal memoirs that aided in this research. I am indebted to Lt. Col. Tom McElmurry, Capt. Wally Schirra, and Cdr. Glenn Tierney for providing copies of flight logs that helped to resolve many questions about dates and details. Aerial transportation to visit Glenn Tierney in Lake Tahoe in uncertain weather conditions was provided by John Swartz and Michael Wright.

I am especially grateful to my friends Frank and Mim Cartwright, at whose house Howard Wilcox and I often stayed while in the China Lake area. In addition to the many insights that discussions with the Cartwrights provided, their home became a base from which investigations were carried out. The unfailing hospitality of Frank and Mim went far beyond the call of duty and provided moral as well as logistic support.

This book did not proceed easily or in a straight line, nor without some heavy sacrifices of time, leisure, and peace of mind. Eastern

Michigan University was kind enough to grant me two sabbaticals, one toward the beginning and another at the end. Otherwise this study has been pursued with my own funds and without official sponsorship. I am also grateful to Mac Greeley, Eldon Runkel, and Anne Gibbons, whose successive editing got the book to a publishable size and cleaned up the many infelicities in the original text.

While every effort has been made to make this book accurate, [xv] inevitably some errors will have crept in. Where they have done so, I take the responsibility for them, as I do for the interpretations offered.

Thousands of people were involved in developing the models of Sidewinder discussed here. To include everyone would have made the book unreadable. Selection was thus a practical necessity. No doubt credit for key contributions also got awarded more often to the people I interviewed rather than those unknown to me. This is one of the risks involved in writing a book of this kind.

Weapons are out of fashion today, though Americans owe much to them. Not all wars are just nor all weapons wise. But our arms have secured the freedoms we hold dear. Without weapons, we would be enslaved. In World War II our triumph over the Axis powers depended on our ability to develop and mass-produce these weapons of war. The Soviet Union, too, crumbled in part because it could not match our weapons. This study shows how such weapons become reality. The men and women who built Sidewinder deserve the thanks and respect of the American people. I salute them and the many others whose honest toil in weapons research and development has kept us from harm.

While the weapons are important, more important is the process of creativity that gives rise to them. I undertook this study because I hoped it would illuminate the creative process. I can only hope that some of the insights I have gained by studying the rich tapestry of China Lake are properly conveyed in the following pages.

A Note on Names

An important concern was what the characters of this drama would be called. In the narratives that formed the principal material for this book, first names or nicknames were typically used. This is how the participants were described to me, and it also reflects the practice in the U.S. Navy, where virtually everyone has a nickname. Thus to preserve the texture of the story, I also have used first names or nicknames to identify individuals in the text. Correct given names are used in the notes. Similarly, military ranks in the text typically reflect the rank at the time the actions happened, while ranks in the notes are the currently held ones. Also note that air force ranks and navy ranks are quite different; a navy "captain" is a far higher ranking person than an air force "captain." I devoted attention to getting titles and spelling correct, but I beg the pardon of anyone inadvertently wronged.

SIDEWINDER

[1]

The Weapon
Nobody Asked For

Research and development is not a business that can be carefully planned and directed, not if you expect to make progress rapidly and economically.

Burton Klein, Economist, 1958

In the Mediterranean

On 19 August 1981 two U.S. Navy F-14 Tomcat fighters from the Sixth Fleet aircraft carrier *Nimitz* were on combat air patrol (CAP) over the Gulf of Sidra. These were waters claimed by Libya under Muammar Khadaffi but not recognized as Libyan under international law. Ostensibly, the fighters were covering a fleet missile exercise—in reality, the exercise had been designed as a provocation to harass and intimidate Khadaffi. On the previous day, U.S. and Libyan aircraft had sparred, testing each other's mettle. The Libyans had not fared very well; on the other hand, the U.S. crews had not learned much.[1]

The crews flying CAP at twenty-two thousand feet on the nineteenth did not expect much to happen. Cdr. Hank Kleeman and his back-seater Lt. Dave Venlet were in the lead; Lt. Larry Muczynski and Lt. Jim Anderson were on the wing. Suddenly, U.S. radars picked up two Libyan Su-22 Fitters taking off and climbing toward the F-14s. The Tomcats soon picked up the Su-22s on their own radar and accelerated to five hundred knots as they closed on the Su-22s; Muczynski slid out into combat spread five thousand feet abeam. The Libyans, obviously under excellent ground control, countered the move, but the F-14s continued to close and were able to identify the Libyan jets.

Kleeman and Venlet were less than two thousand feet from the Libyans when the lead Su-22 fired a missile. Muczynski saw "a tremendous orange flash and smoke trail coming off the [Libyan] plane and going out under Commander Kleeman's plane." The missile then looped upward toward Muczynski but clearly was going to miss him as well. Under the *Nimitz*'s rules of engagement, since the Americans had been attacked, the fight was on.

Both Tomcats carried AIM-9L Sidewinder air-to-air missiles. They were accurate weapons and the navy crews knew how to use them. The F-14s initially turned into the lead Su-22, then Kleeman split and went for the Libyan wingman. In seconds, Kleeman was within five thousand feet of the Su-22 and forty degrees off its tail—well inside the AIM-9L's envelope. With the Sidewinder missile tone growling in his earphones, he fired. Initially, the missile appeared to guide ahead of the Libyan plane, then lagged back and hit it in the tail. Kleeman saw fragments come off the plane. It rolled over; the drag chute deployed; and the Libyan pilot ejected. His parachute opened and he drifted down toward the sea.

Now it was Muczynski's turn. Climbing after the Libyan lead plane, he had no difficulty gaining a firing position; he launched a Sidewinder at the lead Fitter. "The Sidewinder went right up the guy's tailpipe and blew off everything from the wing roots rearward in a tremendous fireball," he said later. Muczynski had to pull hard to avoid hitting the cloud of debris falling from the Libyan plane. Looking down, he saw the remains of the Libyan lead jet spinning toward the water. The pilot had ejected, but it was not clear whether his parachute opened.

The entire engagement, from the initial Libyan firing to the shootdown of the second plane, had taken forty-five seconds.[2]

Weapons That Work

Following that incident, personnel at the Naval Weapons Laboratory in China Lake, California, broke out buttons proclaiming "Navy 2,

Libya 0." China Lakers had developed the Sidewinder, and they were proud of it. Their weapon had worked.

Unfortunately, our weapons do not always work so well. Early in World War II, for example, U.S. Navy aviators and submariners experienced disheartening problems with torpedoes. At the Battle of Midway in June 1942, many courageous torpedo bomber pilots lost their lives lining up their shots—to no avail—against enemy warships. Mark 13 torpedoes properly fired went "everywhere but where [they were] aimed."[3] Torpedoes that did hit often failed to detonate. Many U.S. submarines and crews were lost after the torpedoes they fired missed or failed to detonate, alerting the enemy—who found the submarines and sank them.[4] In the Falklands Conflict of 1982, British ships were hit repeatedly with thousand-pound bombs dropped by the Argentines; every time one of the bombs detonated, a ship was sunk or seriously damaged. Fortunately for the British, most of the bombs failed to detonate; if they had detonated, the war might have gone differently.[5]

Weapons fail for many reasons: bad design, poor reliability, human error. Sometimes the problems emerge during the research and development stage, but often they only emerge later, when hundreds or thousands have been produced.

How are weapons developed? How does a rifle, a missile, or a stealth bomber originate? Are the development processes of the successful and the unsuccessful weapons fundamentally different? Once the shooting starts, the weapon is a fixed quantity: It works; it doesn't work; or it works only in certain situations. But there is an earlier time in every weapon's history when the technology, as Bruno Latour says, is still negotiable, when it is still in the process of becoming.[6]

Can development "thrive on chaos," in the words of management guru Tom Peters? If so, what kind of chaos? What is the inner logic of successful operations, if it is not science? Do we simply throw money at R&D laboratories and let them do what they want? If not, what degree of direction do they need? What accounts for successful innovation in the weapons field? What role do leadership and organizational design play? And what should be the role of government laboratories as opposed to the private sector?

To answer these questions, I focus on a single laboratory, a single person, and a single weapon—China Lake, Bill McLean, and the

Sidewinder air-to-air missile—to show how a key development team came together in a laboratory designed to nurture high creativity.

In the end I move beyond hardware and day-to-day operations to an analysis of the way a creative R&D laboratory works, and beyond the laboratory itself to larger questions of how the navy, or any similar institution, can promote the innovation it needs.

[2]

The Gadgeteer

Weapons systems are human creations. The person who created Sidewinder was Bill McLean. A gifted technologist, he was as remarkable as his missile. Yet the first impression that many people got of Dr. William Burdette McLean gave little hint of his fine intellect. He seemed quiet, shy, and introverted.[1]

His conversations with colleagues often were short on details and context; he touched on aspects of what he was thinking about without delivering a clear picture. An unconventional thinker, he saw the implications of new technologies long before others did; he considered possibilities that simply did not occur to the ordinary scientist or engineer. In the end, he became one of the most respected scientists in the navy. David Potter, at one time assistant secretary of the navy for research and development, called him "the top naval scientist of our time."[2]

He was born 21 May 1914 in Portland, Oregon. His father, the Reverend Robert N. McLean, was a Presbyterian minister, and McLean noted that both his grandfathers also were Presbyterian ministers who had "founded churches in communities where their operations were unpopular." Accepting challenges was a family tradition. So was self-reliance.[3]

The boys learned from their parents. "My mother taught me to knit, crochet, and use the sewing machine before I went to kindergarten. My father showed me how to repair automobiles, build houses, do plumbing and electrical wiring. Everything that broke became a challenge to figure out how to repair it and make it better," Bill said. The family moved often, finally settling in Eagle Rock, California. Bill recalled that "the first priority at each new house was to set up the workbench and the vise. Next came the selection of a place which could be darkened at least at night to serve as a darkroom for developing pictures." For Bill and his father photography was a passionately pursued hobby.[4]

Mechanical things fascinated McLean, and concern for finding the right materials outside of ordinary channels remained with him throughout his life. Years later, when McLean was thinking about the future location of the Naval Undersea Center, Peter Nicol recalls him going to the local hardware stores. In Santa Barbara, McLean observed, "you can't find anything," whereas in San Diego, the stores were better stocked. What influence this observation had on keeping the naval laboratory in San Diego remains unknown.[5]

The family budget was slim. The frugality developed as a child stayed with Bill all his life. One effect was a constant concern with costs—another was a tendency to scrounge, an activity that Bill pursued even after he was well established as technical director at China Lake. It is interesting to speculate what effects his childhood experiences had on the military systems he built. An article in the China Lake newspaper—the *Rocketeer*—announcing Sidewinder's fleet introduction in 1956 would characterize the missile as a logical outgrowth of childhood habits: "This overwhelming urge for simplicity in design, Dr. McLean believes, resulted from the tight purse strings he had to tug against in his youth. If he wanted a photo enlarger or a new electrical gadget he had to design and build it himself. And since economy was a big factor, he had to design for economy in the cost of parts—economy in effort—and economy in respect to the tools he would need."[6]

This is hardly the whole story, but family background certainly was a key influence, though World War II played a major role in convincing him that simple systems were better than complicated ones.

[6]

Caltech

While the McLeans had fun, they managed to put enough aside to send Bill and his two brothers to college: Bill and Jack went to the California Institute of Technology and Bob went to Occidental College.[7]

When Bill McLean graduated from Eagle Rock High School in 1931, Caltech seemed the obvious choice for him. With his intense interest in science and technology, the small but well-run technical university seemed ideal. In fact a more suitable combination of pupil and school can hardly be imagined. McLean got a first-class education in physics and in the conduct of creative research. He would stay for eight years and leave with a doctorate in physics. At Caltech, he encountered a succession of brilliant teachers, supervisors, and peers who helped him refine his talents. Years later, technician George Jennings observed that most of the top technical performers he had met at China Lake and Pasadena seemed to have gone to Caltech.[8]

Caltech offered McLean scientific training of the highest caliber in the company of other similarly minded young men. The school was small, but the faculty was elite and the work was exacting. "You were terrified all the time, but you kept up or left," remembered engineer Rod McClung, who graduated in 1939—the year McLean got his Ph.D. there—and later played a key role in the Sidewinder program as a telemetry specialist.[9]

McLean seemed unfazed by the pressure and took a heavy load. Never a grind, McLean spent hours playing bridge or "spouting radio tube numbers."[10] He carried about fifty-two hours per term, an average load at Caltech, and graduated with a B-plus grade average. Given the standards, this was an outstanding undergraduate record. McClung, for example, ranked ninth in his undergraduate class with a straight B average, which was the lowest grade point average that received an honors degree that year. McLean's Ph.D. was awarded cum laude.

McLean entered intending to become an electrical engineer. After one semester, however, his adviser Earnest Watson suggested "that I might enjoy physics more than engineering because it would provide the more basic fundamental knowledge. The application of fundamentals to engineering he felt could come later. In any case, I saw my engineering friends taking massive amounts of data on the electrical generators in the basement of Throop Hall. They also had to keep extensive notebooks. I transferred to physics and have never regretted the change." Physicists like McLean would play a major role in the Sidewinder project, and many felt that its special character derived from this source.[11]

He graduated in 1935 and immediately entered the Ph.D. program, where he encountered a powerful formative influence: Charles Lauritsen. A brilliant physicist and an experimentalist par excellence, Lauritsen was a student of Robert Millikan, whose influence on Caltech as its head (1921–45) was pervasive.[12] Lauritsen, a former naval architect, invented the quartz-fiber electroscope. This device was an excellent detector of X rays and nuclear particles in the days before electronic instruments were common in laboratories. Scrounging and custom-fabrication activities were important training for Lauritsen's graduate students; he insisted that his Ph.D. students be able to make the experimental apparatus they used.

McLean was in his element. He thrived in this heady atmosphere of constant experimentation, working with Lauritsen and Willy Fowler in nuclear physics, and eventually wrote his dissertation on short-range alpha particles. When Lauritsen later played a key role in the founding of the Naval Ordnance Test Station at China Lake, he would naturally think of enlisting his gadget-oriented student.[13]

McLean enjoyed his work with John Streib on the design and construction of the five-hundred-thousand-volt Van de Graaff generator. When he demonstrated this "lightning machine" to relatives Bob and Claire Blohm, she thought that this was "the thing that he got the biggest kick out of at Caltech." In addition, he helped build the proton accelerator tube for Lauritsen and Fowler. McLean was a natural "gadgeteer."[14]

Some years after McLean had left Caltech, Lauritsen observed to McLean's associate Hack Wilson that, as a student, McLean often seemed more interested in the apparatus used to make the measurements than the measurements themselves. Jennings, who met him many years later at China Lake, put it best: "If you came up with a new algorithm, McLean would sign the patent application, but if you came up with a new device, he would come down to the lab in person to see the hardware."[15]

Two weeks after McLean got his Ph.D. in 1939, he married Edith LaVerne Jones, who had been the family's housekeeper after McLean's mother died. Trained in physical education, LaVerne was a perfect complement for Bill. The two shared an interest in sports and outdoor activities, but their personalities were polar opposites; the woman everyone soon learned to call "LaV" (pronounced Lah-Vee) was an extrovert. Neither high-ranking officers nor difficult situations fazed LaV, and her memory for names was phenomenal. After years of separation from a casual acquaintance, she could remember all the children by name. On more than one occasion, her ability to remember the name and interests of an admiral to whom she had been introduced only once helped promote Bill's projects. Bill, by contrast, often forgot names.[16]

[9]

Many years later, when Bill was the technical director at China Lake, LaV kept in close touch with the station's human heartbeat. In addition, her outstanding qualities as a hostess led one friend to describe her as "the Perle Mesta of China Lake." Few aspects of the lives and concerns of her community escaped her attention. McLean picked the right university, and he certainly picked the ideal partner for life.[17]

In September 1939, as war broke out in Europe, a postdoctoral fellowship took McLean to the State University of Iowa, where he thrived under another brilliant teacher, Alexander Ellett, doing research work in low-energy nuclear physics. McLean also supervised graduate student research, and when Ellett was called to Washington, took over his class on advanced theoretical physics. But he disliked teaching and was not good at it; he had difficulty explaining abstruse concepts to his students. Henry Swift, one of his pupils at this time and later a close friend, recalls the mutual frustration of teacher and student. McLean was often several jumps ahead of where others started, and he did not always fill in the gaps. This trait quickly separated those who spoke with him into those who could follow his thoughts and those who couldn't.[18]

World War II swept up McLean and his generation. He put nuclear physics aside to spend the next four years in Washington in a government laboratory designing weapons. Ellett, located at the National Bureau of Standards, was head of Division IV—Ordnance Accessories—of the Office of Scientific Research and Development (OSRD), the government's scientific brain trust, and he arranged for McLean to work at the bureau in July 1941.[19]

The Bureau of Standards

At the bureau McLean worked mostly on proximity fuzes for rockets and bombs, while a similar group under Merle Tuve at Johns Hopkins

University was working on fuzes for projectiles, a more difficult design job. Engineer Jack Rabinow put it simply: Johns Hopkins was working on fuzes for things that rotated, and the bureau people were working on fuzes for things that didn't. When Henry Swift first saw McLean at the bureau, he and Rabinow were molding bakelite propellers for the proximity fuzes. "Neither Bill nor Jack knew much about this subject, but they just went right ahead and did it."[20]

Rabinow, a Russian immigrant, vividly recalls his first meeting with McLean. In 1941 Rabinow was working at the bureau as a mechanical engineer. His unusual mechanical talent was already evident in his designs. Although Rabinow was contributing, he felt he was just marking time until he was drafted. Rabinow's supervisor, Hugh Dryden, had other plans for him.[21]

A man of few words, Dryden asked him what he knew about ordnance, to which Rabinow replied, "Not very much." Dryden handed him a book on the subject and told him to read it. Rabinow read the entire volume over the weekend. When he reported back, Dryden mentioned "a few problems" and asked him to design an air pressure–actuated switch. Rabinow designed the switch and asked the machine shop to make it. The technicians knew that Rabinow was a superb technician—he cut the threads for his Leica's telephoto lens by hand—and were eager to help. They were impressed by the fast-talking engineer who could also use machine tools. "I could get the machine shop to do anything," he recalls, "because anybody who could cut threads by hand was God." They made the switch overnight and Dryden was duly impressed. When Rabinow carried off a second design job with equal aplomb, Dryden had seen enough. "Come," he said. "Where are we going?" Rabinow asked. "Never mind, just come."

Dryden took him to a room in the next building. Full of tinkerers, the room seemed a madhouse. Dryden, unperturbed, called to a young man who turned out to be Bill McLean. Dryden's introduction was typically brief. "Bill, this is Jake, he has mechanical aptitude." That was all he said, but from Dryden, this was high praise. When Rabinow asked what they did, McLean merely replied, "You'll find out." Thus Rabinow joined McLean's group specializing in bomb and rocket arming systems.[22]

Inventor Jacob Rabinow about 1950 with instant-reverse motor. Courtesy of Jacob Rabinow

William Burdette McLean, technical director of Naval Ordnance Test Station, China Lake, 1956. Courtesy of China Lake Naval Weapons Center

Each quickly learned to appreciate the other. Rabinow, a natural inventor, described McLean in turn as "the best engineer I ever knew." Rabinow found McLean capable of deep understanding, yet able to simplify complicated matters. "He was very direct and simple about his physics"—and his designs. As with Rabinow, the technicians respected McLean for his mechanical abilities.

McLean had little patience with stupidity. Rabinow recalled a meeting when "very silly things" were proposed. Surprised by McLean's silence during the meeting, he confronted him: "You know what they said in there was bullshit. Why didn't you say anything?" Replied McLean: "Oh, I'm going to do it my way anyway." And he did.

McLean's operation started with ten assistants. Between November 1942 and January 1944 he was promoted twice, and his original group of ten grew into forty people engaged primarily in the development of aviation ordnance. At the bureau, people saw what needed to be done and did it. Bureaucratic matters were largely ignored. Rabinow, for instance, worked in the group McLean headed but was unaware at the time that it had a name. (It was "Mechanical Design," and Rabinow later became its head). "Who's the boss?" Rabinow inquired of Dryden one day. "Don't ask stupid questions," Dryden replied.[23]

While they were learning how to run an R&D operation, Rabinow and McLean were evolving a style that later became identified with the Lockheed "Skunk Works," but actually such informal operation was typical of OSRD laboratories in World War II.[24] Rabinow, who has a gift for turning a phrase, over the years evolved a set of laws:

[Law] #13 I think, says that everything you do illegally, you do efficiently. This, of course, is perfectly obvious. For one thing, you do not write at all because writing on an illegal project is suicide. For another thing, you work with whatever equipment you have on hand, and of course, you do everything on your lunch hour, which starts at 8:00 in the morning and finishes at 5:00 in the evening. Another thing, when it doesn't work well and it is illegal, you drop it very quickly and kill the project. When it is legal, you carry it on to doomsday, hoping that somebody else will carry it on, so that when it finally fails you won't be blamed. If an illegal project does succeed, you will be a hero, but if it fails you would like no one to know about it, so you bury it quickly. Illegal projects are very, very efficient from many points of view. We were allowed to do much of this.[25]

McLean and Rabinow agreed on an experimental approach to design. On one occasion the two men were working on a parachute opener for the Bat antiship missile, which saw service late in the war; it weighed more than a ton, and test versions required a big parachute for recovery. The two decided to use a black-powder device to open the parachute and guessed at the correct charge. It exploded when they tested it. They reduced the charge and it worked. Proud of their work, they marched into Dryden's office and fired it off again. People burst in wondering who had shot Dryden, who said, "Don't worry about it, it's only Bill and Jake showing off."[26]

[13]

Today Rabinow, a professional inventor with more than 230 U.S. patents, works as an adviser to a National Institute of Standards and Technology program on "energy-related inventions." All inventors, he says, come up with lots of ideas, but the smart ones know that most of them are trash. McLean had many ideas, but he quickly discarded the trash.[27]

Rabinow recalls: "On another occasion, the issue was underwater versus surface firings of ballistic missiles. Three out of the four people said you should fire from the surface, and McLean said you should fire from underwater. If you surface, the enemy will spot you; also, if you surface, the ship will roll; if you fire from underwater, the platform will be stable. The whole committee was against him. But Bill McLean didn't care. He was like that."

What McLean cared about was getting the right answers. He spent hours pondering designs and trade-offs, which could put others in the conversation at some disadvantage since McLean might start in on something he had thought through completely, while the others had just started.[28] He often seemed oblivious to the external world while pondering. Once, called to dinner, he went to the table, ate, was somewhat quiet during dinner, and went back to the couch. Suddenly, he seemed to wake up, whereupon he threw up his hands and said, "I've got it! When do we eat?"[29]

At the bureau McLean was learning more than just how to get things done. He was learning the perils of designing complicated items that then had to be manufactured. Working on fuzes provided an intensive education in the realities of production:

> I believe this early experience in the design of fuzes was the most valuable training I have ever received. . . . It is unfortunate that every designer of military equipment cannot at some time be exposed to the problems of designing fuzes and their arming mechanisms. [It] requires the most rigorous attention imaginable to a multitude of simultaneous design requirements. A fuze must be designed with a minimum of parts, with each part doing a multitude of functions. Its reliability [must] be above the 90 percent mark, and its probability of failure in an unsafe manner should be vanishingly small. Many hours of effort at the design board and in testing must be spent trying to design the various pieces in such a manner that it is humanly impossible to assemble them in an unsafe position.
>
> I learned an unforgettable lesson about the difficulty of designing military equipment when I visited the assembly line for one of our new fuzes. One part . . . included a boss on a rotor which was carefully located so as to prevent the insertion of the part in the armed position. On the production line, however, this part appeared to be superfluous and sometimes in the way. It was, therefore, being carefully removed by filing as one step in the production procedure. I have never again been tempted to believe that a product can be produced by means of drawings alone.[30]

At the bureau, McLean became involved in two projects that later affected his Sidewinder work. The first was the Bat antiship missile, a large, slow, bomber-launched glide bomb with two versions: one guided remotely by television and a second with an onboard radar seeker. The missile got McLean thinking about guidance in general and in particular about transferring the fire control system from the launching aircraft to the missile itself. The second was toss-bombing, in which a shrapnel bomb was lobbed from a fighter to explode in the midst of an enemy bomber formation. Bombers were already of concern, and the concern increased after the first atomic bombs were dropped. Toss-bombing would bring McLean to China Lake, where he would figure out that guided missiles, not toss-bombing, were the best way to bring down bombers.

McLean had found good teachers. His formative experiences gave him a deep grasp of science and a practical attitude toward design. He found his niche at the Bureau of Standards and honed his design talents under the pressure of the war. He now moved into an environment designed to give these talents full expression.

The Problem Takes Shape

Can do!—Seabee Motto, World War II

The Naval Ordnance Test Station

World War II made it clear that pure scientists can invent practical things when they have to. "Can do!" applied not only to the Seabees but also to the Office of Scientific Research and Development (OSRD), which developed many of the American weapons used in the war.[1] OSRD was the brainchild of Vannevar Bush, engineer, inventor, and statesman of science, who thought the country needed a special organization to harness its technical talent and quickly organized a group of advisers for the president. Bush's original small advisory group became the National Defense Research Council (NDRC), which in turn evolved into the large and powerful OSRD. Bush recalled: "There were those who protested that the action of setting up N.D.R.C. was an end run, a grab by which a small company of scientists and engineers acting outside of established channels got hold of the authority and money for the program of developing new weapons. That, in fact, is exactly what it was."[2]

OSRD teams shaped most of the government's research activities during World War II, and the unusual organization fleshed out Bush's vision of what science and technology could do. Most of the war's major weapons, from the proximity fuze to the atomic bomb, were initiated, developed, or improved by research done through OSRD-sponsored programs.[3] A scientist who worked on the proximity fuze at the Applied Physics Laboratory of the Johns Hopkins University described the process this way:

> Even within the laboratory, the minimum of organization and hier-archy was permitted. The fluidity of personnel was characteristic of the situation. Each man did the job for which he was best fitted. One day he would be directing his associates. The next day he would be one of the associates under the direction of the man he was directing yesterday. This was true in all stages of development and production, with the organization itself changing as needed. A man in charge of a project might find himself replaced by a notice on the bulletin board setting up a new organization, signed by a characteristic "This is it!—M. A. Tuve."[4]

[17]

The most important invention of the war was OSRD itself. Fortunately, a few wise leaders realized that its style of operation and habits of thought would be needed in the postwar world, and they kept the spirit alive.

Some in private industry learned the same lesson. The Lockheed Advanced Development Center, commonly known as the "Skunk Works," was the archetype. Started in 1943 to develop one of the first jet fighters, the original Skunk Works rose out of packing crates and an old circus tent next to Lockheed Corporation's Burbank, California, wind tunnel. Director Clarence "Kelly" Johnson had promised the air force to develop the P-80 Shooting Star in 180 days; his hand-picked team did it in 141. It was not part of OSRD, but it played by the same rules.[5]

OSRD was disbanded when the war ended, but the Skunk Works lived on to create a series of spectacular aircraft—the U-2 and SR-71 spy planes, the C-130 Hercules transport, and the F-117 "Stealth" fighter represent aircraft that were produced in record time at low cost. The name "skunk works" has since become a generic term applied to streamlined R&D teams allowed to disregard corporate rules to get things done.[6]

McLean learned the OSRD style at the Bureau of Standards and he never forgot it. When he arrived at China Lake, he found the same principles at work. But while the Skunk Works was small, China Lake had thousands of employees. How could skunk works principles be scaled up to an operation the size of China Lake?

The Rocketeers

Even before the war, men with foresight had seized on what became the OSRD style. Rear Adm. William S. "Deak" Parsons, U.S. Navy, and L. T. E. Thompson had served as experimental officer and chief scientist, respectively, at the Naval Proving Ground at Dahlgren, Virginia, in the 1930s and dreamed of establishing an integrated navy R&D center.

> Parsons and Thompson were an exemplary military-civilian team, and what emerged from their years of association at Dahlgren was a common philosophy for effective weapon research and development. . . . The evidence is clear that each man had a profound effect on the other and that the foundations for this went back to their Dahlgren association. After Parsons had risen to prominence in the defense establishment, it was reported of an interview with him that [Parsons] modestly gives some of the credit to Dr. [Louis] Ten Eyck Thompson of the Dahlgren Naval Proving Ground, who, he says started him on the right road back in 1930.[7]

Parsons and Thompson discussed their frustrations with navy R&D and foresaw the need for a naval laboratory that would operate on a different set of principles. (Vannevar Bush later said he gave the atomic bomb project to the army rather than the navy because of the service's shortcomings in military/civilian-scientist relations.) Because they chafed against official restrictions before the war, both men sought to maintain OSRD's principles after it ended; both played powerful roles in shaping China Lake—Thompson on the inside and Parsons on the outside.[8]

[18]   What became the Naval Ordnance Test Station (NOTS), China Lake, California—in the Mojave desert north of Los Angeles—began in 1943. NOTS was a testing range and laboratory for rockets being developed by Caltech in a program begun by Charles Lauritsen, McLean's mentor. Lauritsen had worked at OSRD early in the war, and though his tenure was brief, he appreciated that the armed forces needed good rockets and he wanted to develop them. OSRD had a

program underway at Indian Head, Maryland, but when Lauritsen found its directors unwilling to respond to his concerns he began lobbying for a rocket program at Caltech. He succeeded, and the Caltech program got under way on 1 September 1941.[9]

Testing rockets near Caltech posed problems. There simply was not enough room, and air-to-air firings would demand even more space. After a fizzle near Goldstone, California, a space in the Mojave Desert near the tiny settlement of Inyokern was mapped out in 1943 as the site of the Naval Ordnance Test Station.[10]

[19]

Lauritsen soon turned the day-to-day work over to his former student Willy Fowler (then an associate professor) and a young graduate student Conway Snyder. Design and testing soon produced some successful concepts; when they were mature, they were produced in quantity, either at China Lake or elsewhere, and shipped to the various theaters of war. One of these successes, the five-inch "Holy Moses" air-to-ground rocket, later supplied the motor tube for the Sidewinder.[11]

China Lake began as the solution to an immediate problem: finding space to develop and test rockets. Its functions, however, soon expanded into those of a full-fledged R&D laboratory. During World War II it proved its worth under the joint direction of Caltech and the U.S. Navy. While its character was formed in the crucible of wartime experience, it owed much to the lessons learned by Thompson, Parsons, and others before the war. Lauritsen applied these when establishing the guidelines for the new R&D center—as did China Lake's first commander, Capt. Sherman E. Burroughs Jr., U.S. Navy. Close military-civilian cooperation was a cornerstone.

After the war, the vision of Lauritsen and Burroughs got crucial support from navy leaders who encouraged China Lake and the Naval Ordnance Laboratory at White Oak, Maryland, just outside Washington, D.C., to maintain the free-wheeling, wartime OSRD culture—not an easy thing for a government installation. Weapons firms had boomed during the war and wanted orders to continue; many felt that private industry should be developing the weapons of the future and that the government's role should be limited to testing and evaluation.[12]

China Lake struggled against this trend. While Lauritsen and Burroughs got along well, commanding officers were replaced on normal rotation, and Capt. James B. Sykes, who succeeded Burroughs, felt the installation should run "like a battleship," with the commander clearly in charge. Responding to Sykes, the China Lakers drafted their own principles of operation. These were crafted by many hands but mainly by L. T. E. Thompson, the new technical director, and by Cdr. John T. "Chick" Hayward, the experimental officer. Sykes received the principles with some reserve. On 21 October 1946, however, Rear Adm. George F. Hussey, head of the Bureau of Ordnance, approved them. Prominent in the document was this statement: "It is the intention of this station that its facilities will be so organized and operated as to insure the successful conduct of its research, development, and test program fully equivalent to that attained during the war by the O.S.R.D. groups working in the corresponding fields."[13]

Later, support came from Adm. M. F. Schoeffel, head of the Bureau of Ordnance, on 22 June 1951:

At these two stations [China Lake and White Oak] the Bureau has since 1946 been engaged in an experiment in the method of operating large scale military laboratories. At that time the Bureau decided to operate these installations on the principle that the technical activities would be conducted and directed by professional civilian scientific and engineering personnel, and that the role of the military personnel would be that of providing the necessary knowledge of operating conditions plus the administration required to make the laboratory a part of the Naval establishment in the broadest sense. With this in mind those laboratories have consistently been staffed with professional civil service personnel of the highest quality obtainable, under the leadership of a Technical Director in whose hands the responsibility for the technical achievements of the laboratory is placed.[14]

[20] At China Lake, the partnership between the military and civilian elements continued. While the base had a naval commander, civilians carried out research under a civilian technical director. Uniformed personnel, however, were critical to success. They represented the customer, taking their direction from an experimental officer who was responsible for supplying officers and enlisted men to test the bombs, rockets, and fire control systems. They brought fleet experience to

L. T. E. Thompson, technical director of Naval Ordnance Test Station, China Lake, 1951. Courtesy of China Lake Naval Weapons Center

Michelson Laboratory at China Lake, 1959. Courtesy of China Lake Naval Weapons Center

weapons evaluation and acted as advocates for the systems after development. The subsequent careers of some experimental officers are testimony to the value of training received in these tasks and the quality of men chosen to do them. Among them were Adm. Thomas Moorer, who became chairman of the Joint Chiefs of Staff, Vice Adm. William Moran, Vice Adm. Thomas Walker, Vice Adm. Thomas Connolly, and Vice Adm. John T. Hayward.[15]

As China Lake grew into a full-fledged laboratory, its virtues became clear. The siting of laboratories together with the testing ranges might seem like mere convenience but in fact was crucial. Novel ideas could be developed and tested with little paperwork or delay. The key was to employ the full spectrum of R&D activities from initial conception to pilot production. Only after NOTS developed many important systems, such as Sidewinder, was the value of the arrangement made clear.[16]

A Shaped Charge

World War II had proved to many that delay was the enemy. Merle A. Tuve, the maestro in charge of developing the proximity fuze, once said: "I don't want any damn fool in this laboratory to save money. I only want him to save time." Those who had worked in OSRD during World War II understood this point.[17]

China Lake had specialized in rockets during the war. When a problem developed, solutions were quickly sketched, translated into prototypes, tested and revised, then released for production to the fleet or the army air forces. China Lake traits became famous: rapid response, high sophistication, and user-friendliness.

Early in the Korean War, Soviet-designed and -built tanks threatened to push the United Nations forces completely off the peninsula. China Lake was ordered to develop antitank rockets capable of piercing 11-inch armor plate (a demanding—and ultimately unnecessary—specification discussed more fully in chapter 7). Scientist Emory Ellis took charge, decided that a shaped-charge warhead was the answer, and directed the design of the 6.5-inch antitank aircraft rocket RAM in a crash program that operated around the clock.

This weapon went into production on the station. Fuzes were assembled from locally available materials such as clothes-pin springs and hearing-aid batteries. The assembly line for the first units ran down the main corridor of the Michelson Laboratory, often called simply the Mike Lab. Secretaries and wives came in to work on the line after hours. Safety was stretched to the limit, but twenty-nine days after Washington had issued the order, the first thirty-five pilot rounds were en route to Korea. Mass production followed.[18]

China Lake culture also drew on its Caltech origins, where the lights were always on for those who wanted to work.

[23]

Finally, L. T. E. Thompson, the new technical director, provided the guiding spirit to animate the base and its civilian personnel. An elegant and accomplished scientist, his priorities were recruiting and the welfare of his people. He set the tone for the complex dynamics of the laboratory and the base community. Technically gifted, he was at the same time the quintessential people person. When he was putting together his advisory board, Admiral Hussey, then chief of the Bureau of Ordnance, told him he would never get the people he wanted. But Thompson's reputation was such that he managed to get most of them.[19]

In building a staff, Thompson naturally sought out gifted scientists like McLean. The esteem in which McLean was held even in 1945 may be inferred from the two men who came to recruit him. Capt. Kenneth H. Noble, deputy director of the Bureau of Ordnance, and Thompson himself visited McLean at his home in Washington, D.C., and asked him to drop by China Lake. The stated reason for the visit would be to try out toss-bombing techniques, which McLean was working on at the bureau. Noble and Thompson, however, were thinking about a permanent move. They pointed out that China Lake would be exciting in the postwar era. McLean knew Thompson from wartime work at the Naval Ordnance Factory in Indianapolis. He knew that Lauritsen and Fowler were closely connected with China Lake.[20]

But congenial colleagues were not the only attraction. Equally important were the easily accessible testing ranges. At the Bureau of Standards, testing had been done at distant ranges; going to the range meant, at the least, a two- to three-hour trip by car. "So the very appealing thing at China Lake," McLean said, "when I first went out there, was the fact that the laboratory was planned to be close to the test range. You could work on your equipment and then take it out and get it tested and come back and work on new elements of the problem."[21]

McLean went out to California at Easter 1945. He liked what he saw and came back in July with his family, intending to stay three months. When he left in 1967, he was China Lake's technical director, its most famous scientist—and he had helped fulfill the hopes of its founders.

In 1945, however, China Lake was not much to look at. Not a tree was in sight. Most of the military personnel and civilians who worked at the base lived there, but housing was hard to find. Civilization (Los Angeles) was four hours away. The neighboring town of Ridgecrest had few stores, and the base commissary was not well stocked. Many items had to be ordered from the Sears and Roebuck catalog. Some people arrived, took one look, and left. But McLean liked what he saw. He could see the possibilities.

Testing Missiles

China Lake moved early into testing guided missiles because of its expertise and facilities. Along with the Naval Air Missile Test Center at Point Mugu on the coast, China Lake's initial role was to test missiles developed somewhere else. What turned out to be its most famous product, however—the Sidewinder—did not evolve from an outside assignment. It came, unbidden, from within.

The base was well equipped and its Michelson Laboratory, which opened in 1948, was first-rate. Nevertheless, projects often outstripped their funding. Scrounging from outside sources and salvaging materials at hand was standard practice. But the base had one overriding asset: the ability to take a system from initial conception right on up through pilot production, all in one place. For McLean, this "full-spectrum" character soon became indispensable.

Designing Weapons in the Postwar Era

McLean started as head of the Fire Control Division of the Research and Development Department. This administrative post left little time for original research. Shortly, the Fire Control Division became the Aviation Ordnance Division, and McLean reported almost directly to Thompson, the technical director.

Already, McLean was establishing rapport with like-minded colleagues such as Henry Swift, who had worked on toss-bombing at Iowa State while McLean and Rabinow were laboring at the Bureau of Standards. Swift had heard about China Lake in October 1945 when McLean was on a recruiting tour at Iowa State University and gave a pitch to several interested colleagues and their wives at the Swifts' home. McLean had great credibility with his Iowa State colleagues, and the visit paid off. Several people, including Swift, accepted. Swift and McLean rapidly developed an intellectual synergy. In 1947 Henry Swift countersigned the laboratory notebook for McLean's first recorded sketch of what became the Sidewinder.

[25]

Paperwork was not McLean's strong suit. As his brother Jack observed, "Bill does not delegate responsibility [for paperwork], he just walks away from it." At China Lake, his talented associate Newt Ward picked it up. Ward became McLean's assistant in running Aviation Ordnance when it graduated from division to department status, and McLean became department head. McLean concentrated on technology. Several China Lakers thought that Sidewinder could not have been developed without Ward's help in these matters. Similar partnerships proved important throughout McLean's career.

When McLean arrived at China Lake, his division already had a long roster of responsibilities, but his priorities shifted as he developed a sense of what problems were significant; one of them was shooting down bombers with rockets.

The Bomber Must Not Get Through

As head of the Aviation Ordnance Division, McLean found himself immersed in fire control problems. The most critical one was how to shoot down an enemy (Soviet) bomber armed with atomic weapons. His division was responsible for figuring out how to do it using rocket-armed fighters. In the atomic era, a single bomb could devastate a city. Thus stopping the bomber was the overriding mission. The air force already had a solution: an automated, integrated weapons system. Interceptors armed with sophisticated fire control systems would be guided to attack enemy bombers.

One such integrated weapons system, with Hughes Aircraft fire control, used the F-86D all-weather version of the Sabrejet, armed with twenty-four Mighty Mouse 2.75-inch rockets in a retractable launcher. The first F-86D units were formed in June 1953; by September 1955 the Air Defense Command had twenty wings. Eventually,

Falcon being mounted on plane. Courtesy of National Air and Space Museum

the aircraft were linked to the semiautomatic ground environment (SAGE) system, which provided surveillance of the skies over North America. Other rocket-armed interceptors, such as the F-94 Starfire and the F-89 Scorpion, also were part of this effort, which involved radar, computers, fire control, and interceptors as parts of a vast system.

The Falcon: A Complex Weapons System

Many thought that rockets were not up to the task. Hughes Aircraft had a more advanced system that relied on the Falcon guided missile. Simon Ramo, former vice president of Hughes Aircraft, offered this description:

> The response was a grandiose plan for protecting the skies. The bombers would be tracked by sophisticated radar systems like SAGE, and manned interceptors under computer control (the pilot was along

mainly for the ride) would be sent up to intercept them. The interceptors would lock on to the bombers and radar-guided missiles would kill them. Because the Falcon would be guided by radar . . . it could guide toward bombers regardless of clouds. Since it used radar, it could also attack from any angle, without the need for a tail attack, seemingly the only option with an infrared missile. Tests showed that Falcon really could intercept bombers or even metallized parachutes. The missile would work, and the bomber would not get through.[22]

[27]

The Falcon had mutated to fit the antibomber mission. In March 1946 Hughes had begun developing a small air-to-air missile under an air force study contract. By 1947 Hughes had decided that such a missile might well be launched from the rear turret of a bomber at pursuing fighters; a defense cutback killed the initial project but Hughes continued work on a radar seeker head. As work proceeded, the entire missile program was reinstated, the missile got a name—Falcon—and a dual mission: an antifighter weapon for bombers and an antibomber weapon for fighters. As the Soviet bomber threat increased, however, so did the Falcon's antibomber role. The air force planned to make the Falcon its main air-to-air missile for the 1950s and 1960s.

Hughes seemed up to the job. According to Ramo, it held the largest concentration of top technical talent in the country outside of the huge Bell Laboratories: "Domination of the military high-technology market was Hughes's number one priority." Under his direction, Hughes had absorbed many of the military-systems-oriented superstars from the other companies. Furthermore, unlike many of the other high-technology firms, Hughes had set its most talented teams to work on military systems. Perhaps this is why Hughes won the air force MX 1179 competition in 1950 to develop this radar-guided missile and the radar to go with it.[23]

The air force was impressed by Hughes's technical prowess and was confident the company could produce both the fire control and the weapon for the new system. Gen. Gordon P. Saville, the air force's authority on air defense, had encouraged the contract, and the Falcon was to go into the fighter originally designated "1954," the year both airframe and missile were expected to be finished. The 1954 fighter—which emerged as the F-106 Delta Dart—took much longer to finish than the missile, and an intermediate fighter, the Convair F-102 Delta Dagger, was used instead. The F-106 actually was delivered in 1959.[24]

Falcon Problems

Earlier tests had confirmed that bombers attacking in bad weather posed the most serious test for a defense. None of the late-1940s night fighters could really be considered all-weather, so the air force needed an all-weather fighter and an all-weather missile—and radar was the only guidance system that could work under these conditions.

Despite Hughes's capabilities, the Falcon came with some serious drawbacks. One was complexity: in the radio-tube era of electronics, this meant low reliability. The Falcon had seventy-two radio tubes, not a lot for a "pilotless interceptor," as it was first called, but a lot for a piece of ordnance.[25] (Sidewinder, in contrast, had fourteen radio tubes.) High cost was another; this stemmed from the problems of manufacturing and maintaining a complex missile. Yet another was its small warhead.

Col. Donald Scheller, U.S. Air Force, then a captain, first heard about the Falcon while working at the U.S. Army's Aberdeen Proving Ground, Maryland, with the Optimum Caliber Program before the Korean War. The program's purpose was to determine the key parameters for weapons systems. The Falcon team had come to Aberdeen looking for help in designing a warhead and fuze for its new air-to-air missile. Very quickly, it became evident that the Falcon had a basic problem: its complicated guidance system took up so much volume there was no room for a powerful warhead. While the air force wanted a kill even from a one-hundred-foot miss, calculations showed this would require a three-hundred-pound warhead and a proximity fuze. But the Falcon warhead initially weighed only five pounds and had only a contact fuze.[26]

In fact, the Falcon had to *hit* the target to get a kill. This was not out of the question for Falcon's original antibomber role, but a relatively small, highly maneuverable fighter presented a more difficult target, even when warhead size was increased to forty pounds. The missile's small size remained a difficult handicap.

The F-102 was designed around the Falcon, and the ensuing problems constitute an object lesson in why not to build a weapon system for a particular platform. In the original design, six Falcons were to fit in the missile bay of one F-102. The Falcons could certainly fit, but with the penalties of warhead size just noted.[27]

It is only fair to consider the Falcon's problems against the technological mind-set of the times that focused on hitting bombers. Few foresaw missile-armed fighters engaging in aerial dogfights. Even their designation in accordance with the army-navy (AN) system—AIM for "air intercept missile"—stressed the antibomber mission. (The Falcon became the AIM-4 and the later Sparrow and Sidewinder became the AIM-7 and the AIM-9, respectively.)

[29]

In the end, the Falcon's antibomber role was overtaken by events. The Soviets decided to use intercontinental ballistic missiles (ICBMs), rather than bombers, to deliver atomic warheads. By the time intelligence assessments revealed the Soviet ICBM emphasis in the early 1950s, most air-to-air missile programs were well on their way. The technology was too promising to discard, however, and most such missiles were adapted for use against fighters.

Aiming Rockets

McLean did not set out to design a guided missile; it evolved out of his frustrations over developing aiming devices for unguided air-to-air rockets. Machine guns and even high-rate-of-fire cannons had limited ranges; longer-range rocket-powered weapons seemed the answer. The Germans, often first with technology, had used them during World War II.[28]

McLean began to explore the limits of air-to-air rockets. The available aiming systems stretched the capabilities of both electronics and weapons. For instance, China Lakers Henry Swift and Jack Crawford developed the Mark 16 machine gun and rocket fire control system, a system used in the F-86. It was a rugged, reliable item, but McLean thought such systems were not up to the demands of future aerial combat. The rockets themselves had other problems; unlike bullets, they tended to weathercock into the wind, which made "rocket control more complex."[29]

Because many rockets were needed for an air-to-air salvo, there were stowage problems. The most common answer was a wingtip pod (as on the F-89) or a retractable launching bin in the fuselage (as in the

F-86D). Bob Hoover, the well-known test pilot and virtuoso air show performer, once managed to blow the nose off his F-86D when the bin salvoed the rockets while retracted—the microswitch designed to prevent such an occurrence had been taped over because it had been malfunctioning but no one had told him.

Collision Course

How effective might air-to-air rockets with adequate fire control systems have proved? Instances of air-to-air combat with rockets are hard to find and difficult to evaluate. In a case famous at the time, two F-89s equipped with a total of 208 rockets fired all of them but failed to shoot down an F6F Hellcat drone that had drifted off course and was threatening to crash on Los Angeles. It eventually ran out of fuel and crashed harmlessly. The rockets did more damage. Several started brushfires, and one errant missile hit a pickup truck in the radiator but failed to detonate. Since the pilots were inexperienced, this may not have been a fair test.[30]

Jacques Naviaux, then a Marine Corps F4D Skyray pilot, gave the following rocket-firing recollections, dating from 1961:

> A working radar, which was not all that rare, could detect a fighter at 24 nm [nautical miles], and track it at 20 nm. The plan was to fire a salvo of four 19-shot pods on a 110-degree lead-collision course, with a firing range of 1,500 feet. Whether or not we would have hit anything on a regular basis is a matter for conjecture, but I think not, although I did manage to shoot down a drone for one of the few recorded kills.
>
> We flew practice firing runs using rockets against a Delmar towed target. The firing aircraft contained a Delmar Scorer, a combination radar and a 35-mm camera. Only once while reviewing the Delmar runs did I see both a rocket and a target in the same frame.[31]

[30] All this suggested that rockets would have been poor air-to-air weapons. In two recorded instances, however, air-to-*ground* 5-inch Zuni rockets were used in aerial combat quite effectively. Neither firing aircraft used special air-to-air aiming equipment, and the firing aircraft (both A-4s, one Israeli and one U.S. Navy) used relatively few rockets. Mighty Mouse 2.75-inch rockets would have been about one-eighth the weight of the Zunis, and thus many more could have been carried.

Scheller worked on designs for a short-range air-to-air rocket while at Wright Patterson Air Force Base in the early 1950s. Two F-86Fs were stretched to provide for a belly pod and two cheek pods. Total armament was one hundred small 37-mm rockets; unlike the usual pods, these were designed for wide dispersion. The "shotgun" approach was to be used on a pursuit course. The key, said Scheller, was "short time to target." That meant getting very close.[32]

[31]

The idea is not as wild as it might seem. About half the planes shot down in air combat are shot down by aces, and aces typically score their kills in exactly such surprise attacks—when the enemy is unaware. If we bear in mind that bombers were the original targets for guided-missiles and that the miss rate for early guided missiles was about 90 percent, then perhaps rockets were indeed a useful interim weapon.

The Moving Target

McLean's studies confirmed well in advance what the incidents above suggest: hitting a maneuvering target with a rocket was going to be tough. When realistic measurements of all the factors affecting rocket flight were made, they found they could compensate for all but one: target unpredictability. Even a small amount of target maneuvering would create 20 to 30 mils of error. This conclusion was the key factor in deciding to develop Sidewinder as a guided missile.[33]

To be sure, this finding was not absolute. But the new world of air-to-air combat, thought McLean, made the rocket—and the aerial cannon—obsolete. There would be no more dogfights; the enemy would be shot down before a dogfight could take place. He was wrong about dogfights, but he was right in his conclusion that the advent of guided missiles meant the end of the air-to-air rocket.

Guided Missile Problems

Swift remembers McLean, perhaps as early as 1946, saying that the key was to "put the fire control in the missile, instead of the aircraft," a phrase he repeated many times over the years to anyone who would listen. The statement was not just technically correct; serendipitously, it also justified China Lake's research into guided missiles—because fire control was part of China Lake's mission.[34]

In the late 1940s, however, guided missiles had severe public relations problems, McLean found:

Every time we mentioned the desirability of shifting from unguided rockets to a guided missile, we ran into some variant of the following list of missile deficiencies:

Missiles are prohibitively expensive. It will never be possible to procure them in sufficient quantities for combat use.

Missiles will be impossible to maintain in the field because of their complexity and the tremendous requirements for trained personnel.

Prefiring preparations, such as warm-up time and gain settings required for missiles, are not compatible with the targets of surprise and opportunity which are normally encountered in air-to-air and air-to-ground combat.

Fire control systems required for the launching of missiles are as complex, or more complex, than those required for unguided rockets. No problems are solved by adding a fire control computer in the missile itself.

Guided missiles are too large and cannot be used on existing aircraft. The requirement for special missile aircraft will always result in most of the aircraft firing unguided rockets.[35]

This rhetoric of denial seemed to provide powerful arguments against wasting time on the expensive and complicated guided missile for use in air-to-air combat. Such resistance to new technology is hardly new. An unknown China Laker, who later annotated a McLean article on the Sidewinder, suggested that in a more modern context one could substitute the word "lasers" for "guided missiles."

[32]   It seemed that one could have reliable, low-cost inaccurate rockets— or expensive, highly sophisticated, accurate guided missiles that would be difficult to maintain, prepare, and fire. McLean, however, already was working on two parallel thoughts: One told him that unguided rockets simply would not work; the other, that guided missiles as then conceived would not work either. He decided to reorient his thinking. The ideal solution would be to have the best of both worlds—a rocket

with fire control. The key, he decided, was to use the current problems as a source of design objectives.[36]

To understand the problems associated with the guided missiles of the day, McLean's group visited most of the projects then extant: the former German V-2 scientists at White Sands, New Mexico; the Hermes Group at General Electric; the Sparrow II organization at Sperry Rand; Meteor at the Massachusetts Institute of Technology; Bumblebee at the Johns Hopkins Applied Physics Laboratory; Dove at Eastman Kodak; and Falcon at Hughes Aircraft. "I asked them what their problems were on missiles," McLean said, "then we tried to find solutions to the problems they were having."[37]

[33]

> In looking back over the program, the single most important abstraction concerns the importance of not starting too fast. At the beginning of the Sidewinder program, I personally spent nearly three years on a part-time basis in the process of considering possibilities—mentally arranging them into a missile, checking the tradeoffs, and trying to think of other methods of arrangement which would make the final design more acceptable to the user. At this stage of the development, reorientation of the program is easy. A complete reorganization of the internal workings of the missile can be accomplished in the time that is required to think of it. I believe that this process, by which one man gets fairly clearly in mind a picture of what he would like to produce and the reasons for selecting one set from a multitude of possible choices, is a very important step in the accomplishment of a satisfactory final product.[38]

From this pondering, the answer slowly emerged: the ability to self-guide toward the target must be combined with the simplicity and reliability associated with rockets. Only a missile so designed would meet the objections of the critics.

A License to Think

Management theorist Deborah Dougherty has argued that successful products generally come from companies whose design, manufacturing, sales, and marketing departments focus on meeting the customer's carefully studied and articulated needs. Getting these four departments to focus on a common, well-chosen goal is the secret to success.

What is remarkable about McLean is that he combined in a single person all these activities.[39]

It is hardly surprising that a gadgeteer like McLean focused on simple and elegant designs. He often said, "It is easy to build something complicated; it's hard to build it so that it is simple." This kind of design led in turn to ease of manufacture, a factor whose importance he learned to appreciate during his work on fuzes for the National Bureau of Standards. He wanted to build Sidewinders for one thousand dollars a copy; he expected mass production.

McLean also knew that to sell the missile it had to appeal to its users. The design of the missile itself made marketing easy. The system that McLean evolved, bringing the various aspects of the missile into harmony, was the key to his unique sales proposition. The system would work for its users because McLean knew what they needed; he concentrated on these needs as opposed to bureaucratic specifications—and his approach worked. In time, his team sold Sidewinders not only to the U.S. Navy and Air Force but also to virtually every friendly air force in the world. The missile was so adaptable that later it was installed not only on aircraft but also on helicopters, armored vehicles, and even ships. The "systems" aspect of getting Sidewinder to the field was carefully thought out.

Uniting as he did all these talents, McLean understood that any missile had to be inexpensive and reliable. Other programs did not reflect this philosophy. Even if they achieved their design objectives, their built-in problems doomed them to mediocrity. Fred A. Darwin, then executive secretary of the Guided Missiles Committee of the Department of Defense, put it this way: "Day-by-day, then with increasing acceleration, I became convinced of something I considered important: THESE THINGS WILL NEVER BE OPERATIONALLY USEFUL, Even Should We Make Them Perfect."[40]

[34] In conceiving his missile, McLean determined to use "the minimum amount of garbage you had to hang on the rocket in order to make it home." Experiments would tell him just how much garbage this was. Because the other missile programs had to respond to formal specifications, they could not reorient their activity in this way.[41]

Critics often claimed that China Lake was "just Bill McLean's Hobby Shop" or that its personnel were mere tinkerers who often succeeded through a kind of random trial and error. Nothing could be further from the truth. China Lake developed its goals from a deep understanding of the needs of the fleet, aided by the carrier pilots who served on the station's staff. Developing this understanding took time, but the end result of all the questioning and weighing of potential trade-offs was a sharply defined problem. When *this* problem was solved, the solution was valuable. Given this clear purpose, the China Lakers' willingness to try anything produced not a random drift but a rapid convergence on systems that worked. Former Secretary of Defense Robert McNamara once stated that 75 percent of the air weapons used during the Vietnam War were developed at China Lake—a far cry from the hit-or-miss reputation it had in some circles. [35]

China Lake did not see itself merely as carrying out orders or evaluating systems. Rather, it was a thinking node in the sociotechnical system that was the U.S. Navy. It took pride in its ability to ask fundamental questions and come up with novel solutions. McLean and his team believed they had a *license to think.* As their successes multiplied, this license to think enlarged.[42]

[4]

The Wrong Laboratory

> McLean was very capable of thinking both globally in the
> largest of terms, and in the most minute of terms. If a screw
> had been badly located in a design, it caught his eye. He had
> *design sense.*
>
> *Howie Wilcox*

Moving the Fire Control into the Rocket

McLean was pondering the problems of fire control systems. Projectiles obviously could get closer to their target than launchers, and the closer to the target they got, the more accurate the fire control would be: putting the fire control system in the rocket instead of the aircraft made sense. The key to success, he decided, was to use an infrared (IR) detector—much smaller than a radar. The difficulty would be in getting infrared detection to work.

All objects naturally emit infrared radiation, and the hotter they are, the more they radiate. Jet tailpipes are good emitters. IR was new technology, but McLean was convinced it would work. Fortunately, the navy was in the midst of developing an infrared fuze, which encouraged McLean to use infrared. As Vice Adm. Bill Moran put it, "If you sense the target well enough for the fuze to operate, you should be able to sense it well enough for a little bit of corrected steering at the end of the trajectory."[1]

The initial idea then was to use infrared as a form of *terminal guidance*. The pilot would fire the rocket toward the target, and the "little bit of corrected steering" during the final portion of flight would compensate for fire control errors and target maneuvering.

McLean's idea for an infrared seeker emerged while he was in Boston on business with Firth Pierce, who often acted as his intellectual sounding board.[2] When they got back to China Lake, Newt Ward remembers McLean taking an envelope from his pocket and showing him a small diagram of a "seeker."[3] Shortly thereafter, on 19 November 1947, McLean sketched the idea in his notebook, added a brief description, and had Herbert Hassenfratz and Henry Swift sign it. This signed and countersigned drawing of a "target seeker" in McLean's notebook was the first definite, recorded step toward Sidewinder.

It is important to understand just what he envisioned. The heart of his target seeker was a lead sulfide photocell that could detect infrared radiation. By 1947 it was well known that lead sulfide's electrical resistance decreases when exposed to infrared radiation, and that current running through a lead sulfide photocell will increase under these conditions; the technical term is photoconductivity. Connecting an electrical circuit to a lead sulfide photocell yields a simple infrared detector. McLean might have used a heat-sensitive thermocouple, which had been used during World War II, but a thermocouple would have been too slow.

Detecting the emitted radiation was only the beginning. To make the detector into a missile guidance unit, designers faced two basic problems: The seeker had to follow the target, and it had to communicate guidance information to the missile.

In a drawing and accompanying two-page description, McLean solved problem number one by using a photocell mounted on a rotating gyroscope. The gyroscope included an asymmetrical mirror, which reflected light on the photocell only if the target was off the axis of the missile. The gyroscope relied on electromagnets mounted in a ring around it to create precession and shift the gyroscope's focus toward the target. Signals arriving at the photocell caused the electromagnets to exert a temporary force, always at the same point in the rotation, to

19 November 1947

I discussed target seeker for a homing missile with Firth Pierce yesterday. As a result of the discussion I decided that the following design would make a very compact and lightweight control element.

heat sensitive element.

transparent dome

approx 1½"

parabolic mirror

sintered material light permeability low conductivity

pivot system support.

transmission oil.

amplifier

leads.

alnico rotor two pole.

The alnico rotor can be spun by air pressure and will act as a gyro. The mirror mounted on its top surface will project an image of the target which will describe a circle about the heat sensitive element if the target is centered. If the target is off center the circle will strike the heat sensitive element and generate an A.C. voltage

McLean's notebook diagram of seeker, 19 November 1947. Courtesy United States Naval Air Weapons Center, Weapons Division which can be amplified and used to precess the rotor by interaction between the coil surrounding the rotor and the alnico field. If the position of the mirror is adjusted properly with respect to the alnico rotor the resulting precession will be in such a direction as to turn the gyro axis toward the target. Takeoffs on the position of the gyro can be used to control the orientation of the missile.

A.B. McLean.

Disclosed to and understood by. Herbert H. Hassenfratz
19 Nov. 1947
J.H. Swift
25 Nov. 1947

keep the gyro's axis aligned with the target. The gyro, which could follow the target, became a target seeker.

The gyro solved the second problem by sending similar signals to the missile's steering gear, which deflected the control surfaces to keep the missile on a collision course. The gyroscope would find the target, turn toward it, and then signal the missile to turn itself onto an interception course. Obviously, McLean's seeker would have to be in the nose of the missile, which holds true for air-to-air missiles today.

The gyroscope's role was critical. Air-to-air missiles are relatively small and easily deflected. A missile could pitch, roll, yaw, and oscillate, but the seeker had to stay locked on the target despite all these motions. A gyroscope resists changes of angle and thus tends to retain the orientation of its axis of spin, even as its housing moves. McLean had learned that a gyroscope could be used to stabilize an aiming device when Jack Rabinow had suggested it as a way to stabilize the Bat missile's radar during World War II. The principle was sound. Getting the gyroscope to operate well would prove difficult.[4]

Although crude, the sketch signaled the beginning of research on what McLean would call a "heat homing rocket." The last word is significant, because a rocket was an ordnance item, not a "pilotless interceptor," which accurately described a number of missiles of the time, including the Falcon air-to-air missile.

Also important was McLean's carefully phrased description of what he was doing. From 1946 to 1949, McLean made constant reference to "putting the fire control in the rocket." This provided some bureaucratic protection, since R&D on fire control systems was a legitimate task for China Lake; developing guided missiles was not. Thus McLean's description justified China Lake's attempts to develop it.[5]

Just Experimenting . . .

[40] McLean had glimpsed a potential solution, and he was hooked. He rapidly moved to the second part of his research project: finding out if the key technologies would work. A busy division head himself, he gathered a small group—eight to ten scientists, engineers, and technicians, mostly from the Aviation Ordnance Division—and began to test key components. While McLean was evolving a missile, he was also assembling a team. The heat homing rocket had no official blessing—

and neither did the team. But the project looked interesting, and McLean had a subtle charisma.[6]

The team organized around McLean's vision, which guaranteed commitment when things went wrong, as they often did. Over the next two years, many of the basic ideas for the heat homing rocket evolved from his ongoing dialogue with close colleagues and from Aviation Ordnance Division meetings.[7]

Little was down on paper, yet the small group developed a clear commitment to test whether McLean's infrared seeker would work. This was the critical issue, and Larry Nichols headed the effort. But other issues, such as aerodynamics and fuzing, also needed work; it was a complicated project.

All the while, McLean confronted the ambiguities of his position. He was head of the Aviation Ordnance Division; his concerns were supposed to be guns, rockets, and bombs. He was supposed to be developing aiming mechanisms, not guided missiles. Furthermore, China Lake was a Bureau of Ordnance (BuOrd) laboratory, and it had not been asked to develop an air-to-air missile. In fact, it had been told *not* to do so because it had failed badly on a previous effort using off-the-shelf components. Its NOTS Air Missile had turned out to be a poor design, tarnishing not only the immediate team but the very idea of developing a guided missile at China Lake. McLean was about to commit one of the military's worst sins: developing technology at the wrong laboratory.[8]

Gathering Support

He needed money. Funding is a problem for legitimate projects—and the heat homing rocket was a bastard. Station discretionary funds "left over" from Caltech's rocket program had been turned into exploratory and foundational research funds to get such projects started.[9] In January 1948 BuOrd had chipped in thirteen thousand dollars, but these funds were soon shut off. During 1947 and 1948 the project scrimped by on a considerable amount of volunteer assistance from China Lake scientists and technicians. Even after a formal proposal was made in 1949 for a "heat homing rocket," the bureau made available only one hundred thousand dollars, ostensibly to support work on fuzes.[10] Rear Adm. Malcolm Schoeffel, chief of the bureau, and his deputy Rear

[41]

Adm. Deak Parsons backed the project but could not get their guided missile group to give up any of its own funds. They tapped the fuze group instead, making the argument that the Sidewinder was an "intelligent fuze." McLean noted the rather thin arguments: "So our first support came from the fuze group. That wasn't too bad a stretch of the imagination because the guidance unit really did screw on in the same place that a fuze screws on."[11]

With the fuze support came a high security classification—and low visibility. Fuzes had a higher classification than other kinds of systems, which meant that Sidewinder got little official attention during this period.[12]

China Lake was caught between two bureaucracies—the Bureau of Aeronautics (BuAer) and the Bureau of Ordnance—that were locked in a struggle to control guided missile development that dated back to the 1930s. The two bureaus had different constituencies, and their weapons systems reflected different needs. BuAer served aviators' needs—the "brown shoes"—while BuOrd saw itself as the bureau serving the needs of ships' officers—the "black shoes." But there was ambiguity, overlap, and conflict, and each had poached on the other's territory. The infighting finally had to be settled.[13]

In November 1947 Rear Adm. Daniel V. Gallery, deputy chief of Naval Operations for Guided Missiles, worked out a compromise. BuAer would develop missiles for aviators, that is, those fired from planes; BuOrd, meanwhile, was going to develop missiles fired from ships. The compromise, unfortunately, turned out to be a temporary and unstable truce. The rivalry continued, hindering and also, paradoxically, helping the Sidewinder's progress.[14]

Each bureau soon concentrated on a favored missile program, BuAer on the Sparrow III and BuOrd on the air-defense ramjet missiles represented by the Bumblebee family: Talos, Terrier, and Tartar, for which China Lake provided test ranges. BuOrd was taking its scientific direction from Johns Hopkins University's Applied Physics Laboratory, and McLean found himself competing for resources with the far more powerful Terrier program.

BuOrd laboratories were precluded from developing aviation missiles but were allowed to test them. China Lake, in particular, was told to restrict its research efforts to unguided ordnance, such as rockets and bombs. But since it was an ideal test location, both BuOrd and BuAer tested missiles there until the completion of BuAer's Naval Air Missile Test Center at Point Mugu on the California coast. Thus China Lake carried out early testing on the Lark missile. After Point Mugu was established, however, Lark testing was done there—as it was for the Sparrow family of missiles.[15]

Early Lark and Bumblebee testing taught China Lake a lot about [43] missiles. This testing skill and the equipment that went with it would come in handy in testing China Lake's own missiles. But testing was one thing, development another. Even though there was support for missile research as well as testing from the China Lake leadership, McLean was risking a lot by working on an air-to-air missile at a BuOrd laboratory.

The Competition

Already, other laboratories that were tasked to develop air-to-air missiles were hard at work. While McLean was trying to invent a heat homing rocket, the air force's radar-guided Falcon was well on its way. The Ryan Corporation, famous for its drones, was developing the Firebird missile for the air-to-air role, and the air force's Holloman Guided Missile Test Center in New Mexico was evaluating it.

The navy had been in the missile game during World War II and had several projects underway. At MIT a distinguished group of scientists was working in partnership with Bell Aircraft on the Meteor missile, and the Martin Aircraft Corporation was working on the Oriole missile. BuAer was testing both Oriole and Meteor at Point Mugu, along with three different versions of the Sparrow missile. Sparrow, like Falcon, responded directly to an operational requirement and had official blessings to spare. In fact, in the crowded field of guided missiles, it was singled out along with two other air-defense missiles, Nike and Terrier, for special emphasis by the nation's missile czar, K. T. Keller, who had decided to accelerate their transition to production.[16]

Bert Alton, a navy test engineer, remembered the charged atmosphere at Point Mugu as each of the Sparrow contractors tried out its own missile. Years of effort had gone into preparing each system, and often the tests went badly. Millions of dollars and, in some cases, the fate of whole business divisions rode on the outcomes. For instance, in 1951, when the tests were taking place, Raytheon had about six hundred people at Lab 16 in Bedford, Massachusetts, and seventy-five people working on the West Coast; by 1962, Raytheon would have six thousand people working on the missile. Emotions ran high in the control room, and a man could break into tears when his missile failed.[17]

The Oriole and Meteor dropped by the wayside, leaving the three Sparrow versions in the race. As was later the case with Sidewinder, development work proceeded on three different radar seeker systems, developed by three different contractors. Competition between the Sparrow variants was fierce. At one point, Sparrow II personnel apparently mounted a covert effort to jam the continuous-wave radar guidance of a nearby Sparrow III undergoing ground tests.[18]

The first operational Sparrow was the Sparrow I, a radar beam-rider originally developed by Sperry Gyroscope. Guidance for this missile came from a beam on the firing aircraft. After missile launch, the firing aircraft kept its radar locked on the target, and the missile's control surfaces responded to keep the missile in the center of the beam. The system was tested successfully on 2 April 1951, got its first warhead kill on 18 June 1954, and on 6 April 1955 completed its fleet evaluation. The following year, the Sparrow I—the navy's first operational air-to-air missile—deployed with VF-83 as part of Air Wing Eight onboard the USS *Intrepid* (CV-11).[19]

Sparrow II, the second variant, had an active radar seeker. The onboard guidance-and-control unit sent out its own radar signals and received reflections from the target. While the Bendix Corporation was responsible for the guidance system, a major technical challenge, Douglas Aircraft was the prime contractor and designed the airframe. Sparrow II could not be gotten to work. Some felt the partnership failed because of Douglas's lack of interest, but there were fundamental problems. The active radar seeker was a worthwhile idea, but it was premature in the 1950s, before the more compact solid-state circuitry could be used.[20]

Sparrow III Pulls Ahead

Raytheon's Sparrow III won the contest. Raytheon was working on a semiactive homing device before it got a navy contract. In fact, the

navy had approached Raytheon because it had developed a semiactive seeker, which it had evaluated using the Lark airframe. Since both
Lark and Sparrow were being tested at Point Mugu, it was natural for
Raytheon to put a semiactive seeker in the more advanced Sparrow
airframe.[21]

Raytheon already had achieved an important milestone with the
Lark, which had intercepted an aircraft in flight—the first such success for a self-guided missile. Much of the credit for the guidance system went to Royden C. Sanders, a gifted but sensitive inventor who
had been the guiding genius behind RCA's radio altimeter in World
War II.[22]

[45]

As with Sidewinder, Lark started out with a "tiny handful" of people, but by the time work on Sparrow III began in 1950, Raytheon had
six hundred engineers on the job, directed largely by Sanders. It was
Sanders who proposed to develop the guidance system for the Sparrow III. The navy could not help but be interested in such a proposal
and eventually gave Raytheon a development contract. At this point,
however, Sanders quit. Feeling that Raytheon's higher management
was not backing him, he decided to start his own firm and took many
of the top members of the Sparrow team with him. This nearly sank
the Sparrow III project.[23]

Raytheon persevered. The new team, including future corporate
president Tom Phillips, surprised many by completing the design of
the missile. This team, however, lacked China Lake's spatial and tem-
poral unity. While some seventy Raytheon engineers struggled with
flight testing at Point Mugu on the West Coast, the drawing boards
were located at Lab 16 in Bedford. "The missile effort was conducted
across the width of the continent and kept together by telephone."
Nonetheless, they did it.[24]

Sparrow III's semiactive homing was a compromise between the
"fire-and-forget" capability of active homing and the weight-complexity
benefits of beam-riding. In the Sparrow III guidance was located in the
missile, but the firing aircraft carried the powerful radar that illumi-
nated the target. The target, reflecting radiation from the firing plane's
radar, was tracked by the missile's passive seeker, which put the missile on a collision course. Semiactive radar homing (SARH, pronounced
"Sarah") allowed a smaller seeker, but with it the firing plane had to

keep illuminating the target until the missile hit—no "fire-and-forget" capability.

Although the Sparrow III did well in shoot-offs, its success was amplified by the Raytheon image. Raytheon was a big company, and it did a good job of selling its talent, even to Lt. Cdr. C. C. "Andy" Andrews, the Sparrow test pilot. Some years later, when Andrews was a flag officer, historian Max White asked him why the Sparrow III had won the shoot-offs; apparently, there had been no formal evaluation. Andrews stated that he and his colleagues "just felt that Sparrow III was the best system." But this feeling was influenced by the behavior of the three firms. While Sperry and Douglas were old-line firms, autocratic and bureaucratic respectively, Raytheon's attitude was "How soon do you want it?" This approach won points with the testing personnel. Testing engineer Bert Alton was impressed by the personnel from Raytheon and their methodical and businesslike approach to testing. Raytheon wanted the program, and Jake Leiper, its brilliant public relations man at Point Mugu, made sure navy questions got answered.[25]

The Sparrow III soon replaced Sparrow I as the medium-range fleet defense missile. Blessed and anointed by the naval hierarchy, and produced by a large and powerful electronics firm, Sparrow III has remained in service for decades. The F-4 Phantom II fighter, the workhorse of the fleet, was designed with the Sparrow III as its major air-to-air system. Sidewinders were an afterthought.[26]

The advent of Sparrow III gives a good idea of the barriers that would face the heat homing rocket. Competing against well-funded, officially designated projects meant a struggle to survive. The heat homing rocket would have to demonstrate major advantages; any stumble might well cause BuOrd to decide it was just one missile too many. Bill McLean made sure not to stumble.

[5]

Struggles with Infrared

If we had worked like Hughes and Aerojet, it would have been
like wearing boxing gloves. As it was, we had the whole thing
in the palm of our hands.

Larry Nichols, Optical Engineer

Radar versus Infrared

When Bill McLean decided on infrared guidance, opposition was
guaranteed. If the guided missile community agreed on one thing, it
was that air-to-air missiles needed radar guidance. Only radar could
provide "all-weather homing," a capability written into operational
requirements. A heat homing rocket at best would only be a fair-
weather weapon because infrared could not operate in clouds.

Since radar guidance depended on a microwave source to illuminate the target, it was an active system. Infrared guidance systems, on
the other hand, were passive. They detected heat radiation given off
naturally by all objects. The hotter the object, the stronger the signal.
And since infrared radiation is given off naturally, one has to provide
only a detector.

Infrared detection was new. China Lake personnel were told repeatedly that it would not be developed any time soon, but McLean saw
its advantages. If IR could be made to work, it would require less space.
Unlike semiactive radar homing (SARH), an IR guidance system might
be completely contained in the missile itself, while SARH always

would require a target radar-illuminated by the attacking aircraft. The attacker would be forced to close on the target while the missile was in flight. By contrast, since IR missiles carried their whole fire control system on board, the firing aircraft could depart as soon as the missile was fired. IR would give "fire-and-forget" capability.

Studies on Infrared

There were indeed serious problems with infrared guidance. The Germans had failed in their attempts to develop IR during World War II. Engineer Franklin Offner had used a sensitive thermocouple in a U.S. Army infrared guided bomb; his GB-6 had worked, but the army was not interested. And Eastman Kodak used infrared guidance on the Dove missile, a guided bomb being developed by Polaroid. More important, the Dove used an infrared detector with a rotating gyro and a coil that used signals from the detector to precess the coil. McLean had visited Eastman Kodak where the Dove was being developed. Dr. Harold Hammar, a Kodak scientist, was responsible for the IR guidance and it was probably Hammar who first got McLean interested in the subject.[1]

Dove turned out to be a bust; the only round to hit the target was the one on which the testers had forgotten to remove the lens cap.[2] While the Dove's infrared detector worked well, the mechanical design of the missile's seeker left much to be desired. McLean later felt that the Dove's problems had given infrared guidance a bad name. Eastman Kodak was also developing infrared for use in a proximity fuze. But since the work on this fuze was highly classified, it did little to shape overall scientific opinion on infrared. Scientists generally considered infrared technology beyond the state of the art. Basing a seeker on infrared was a big risk. Undeterred, McLean decided to whittle down the uncertainties through research.[3]

[48]   ### Into the Red Zone

Realizing that his Aviation Ordnance Division would need expert help, McLean turned to China Lake's Research Department, particularly Roger Estey, head of the Applied Science Division and an expert in optics. Estey and others were involved in a long-term "target survey" program that L. T. E. Thompson had initiated in 1947. In 1948 Estey assigned Larry Nichols, in the Optics Division, to studies that were to last for many years; Nichols was responsible for getting infrared detection to work—a tall order.[4]

Nichols's first problem was to design a device that would detect anything at all. In 1948, after some crude tests of the lead sulfide cell, Nichols assembled a crude "radiometer" to determine how well it could pick out a jet tailpipe against clouds and clutter. China Lake didn't have any jets at the time, so Nichols mounted the apparatus in the turret of a prop-driven TBM Avenger torpedo bomber and flew to nearby Edwards Air Force Base where he begged air force pilots to fly their FJ-1 jets as targets. Several proved willing; after all, flight time is flight time. Things were simpler in those days.

While the TBM flew straight and level, the air force jets flew by and Nichols pointed the detector at them. The signal received was displayed on an oscilloscope and a camera panned back and forth between the oscilloscope screen and the jet. The size of the jet in the photo was compared against the radiometer reading to determine signal strength versus distance to target.

Early results showed that a detector could work, but only if it had greater sensitivity, thus greater range. Nichols's detector could pick up a jet exhaust at one hundred meters, but not at two hundred meters—and they needed something like two miles. Work continued, and the lead-sulfide photocell's detection capability improved, although it could not track. To become a seeker, it needed to *follow* the bright source of radiation as the target moved.

As Nichols struggled to develop a more powerful detector, McLean began to work on the seeker that would use it. Already in February 1948 engineer Don Duckworth had used McLean's original sketch to build a tracking device. This device had a slanted mirror mounted on top of a rotating gyroscope. The gyro was turned by compressed air, and electromagnets were used to precess the gyro, keeping it pointed at the infrared-emitting target. This primitive seeker could track a light bulb, but at best it was a lab curiosity, too fragile to be put on a missile. In fact it would not work on a missile. So McLean encouraged several small groups to work on more robust trackers.[5]

By mid-1951, however, the project was stalled. Work on the trackers proceeded, but McLean needed a convincing demonstration beyond the confines of the laboratory. None of the seekers was ready to mount on a rocket or ready even for flight tests using "captive carry." At about this time, he or someone else in the group had a brainstorm: Mount a detector on a radar antenna and use feedback from the IR detector to get the antenna to follow the target. The antenna would thus become the "seeker." An old surplus SCR-584 radar pedestal would do the job. It had servomechanisms that would allow it to follow a target, and it was free. If it worked, it would answer many questions and would provide visible proof that an IR seeker was feasible.[6]

Nichols got to work, with aid from Rod McClung, an electrical engineer who by then had joined the Sidewinder team. McClung sought an amplifier with automatic gain control (no problem) that could handle very powerful signals (big problem). McClung's partner, Dave Simmons, came up with the right design and it was incorporated into the pedestal.[7]

Nichols also got help from summer intern Jon Mathews, "a brilliant kid with a mind like a computer." Mathews later took his Ph.D. at Caltech at age twenty-four and immediately joined its faculty. Even at eighteen, however, his mathematical skills were impressive. Rewiring the detector was complicated, but Mathews persisted. McClung and Bob Hummer helped in wiring up the pedestal itself; how it worked was mysterious to Larry Nichols. Mathews concentrated on getting the detector connections wired properly. In November 1951, while a visitor from the Test Department was repeating predictions that "the thing would never work," the antenna began tracking aircraft. Nichols was stunned: "I couldn't believe it!"

The SCR-584 pedestal immediately became not only a critical test instrument but also an unparalleled marketing tool, and a second detector was soon mounted on the pedestal. The two heads permitted comparisons of the performances of different components of the optical system, such as reticles and filters. An antenna-mounted camera showed what the detector was tracking. The tracker was visible proof that an IR seeker could track a bright object automatically, something that had not been demonstrated before. It tracked lighted candles, birds, and even bugs. Crowds came to committee meetings just to watch the tracking films. The heat homing rocket began to look a lot more feasible. When real seekers were developed, they were checked using similar ground test techniques. In 1952 the team mounted two different seekers on tripods on the roof of the Michelson Laboratory. The B version seeker, under the direction of Luke Biberman, tracked better than the A head developed by Avion.[8]

Such proofs were enormously helpful to McLean, who was constantly challenged by visitors from Washington wary of what they called expensive, complicated missiles. Many doubted that IR could work at all—until the old SCR-584 antenna tracking lighted candles made believers out of them. During the critical visit by the Department of Defense's R&D Board in November 1951, the 584 proved its worth. Film still wet from the developing tank was brought into the meeting to provide convincing proof of concept.

[51]

Cloud Worship

But the detector could be fooled. Bright clouds gave it serious problems. Indeed, the tracker initially preferred tracking a bright cloud to an airplane exhaust. As early as 1950, designing a system that would discriminate between planes and clouds had been a major objective, and the answer appeared to be a small device called a "chopper" reticle that broke down the incoming signal. The reticle, which spun with the gyro, was a pattern mounted on a tiny disc that filtered the radiation arriving at the photocell. With some sectors on the disc open and others closed the reticle alternately passed and obscured the beam of infrared radiation, hence the term "chopper." While large sources provided a continuous tone, the chopper turned point sources such as planes into an on-off signal, as each of the spokes alternately obscured or revealed the target. The system was then tuned to ignore continuous tones, thus avoiding larger sources. The initial reticles used film emulsions, but this was eventually replaced by aluminum patterns on glass.[9]

The team tried various speeds for spinning the optical system. They finally settled on 70 cycles per second (cps), the speed for production versions. A frequency meter, consisting of a row of steel reeds, would be held next to the seeker during final checkout, and the frequency would be noted; this was an important test.[10] Nichols notes that "the basic reticle pattern contained six opaque and six transparent sectors in 180 [degrees] with the remaining 180 with no pattern. When a target image was 'chopped' or interrupted by this rotating pattern, it was modulated into an electrical wave form that contained a 'carrier' at 1,200 cycles per second with sidebands at 1,100 cps and 1,300 cps. The carrier indicated the presence of a target and the sidebands gave direction information." Everyone seems to have a different idea about the rates at which the optical system spun—the confusion, no doubt, caused by the various prototypes. For the later AIM-9D it was necessary to increase the spin to 125 cycles to get enough angular momentum. Subsequent production models also used 125 cycles.

The team tried many reticle patterns. The one that seemed to work best was a checkerboard pattern, for which Luke Biberman received a patent. Wilcox, however, remembers the invention of the reticle as having taken place in the course of a struggle between China Lake and General Tire and Rubber (later Aerojet) over who would develop a heat-seeking missile—a struggle that China Lake nearly lost.[11]

Osborne and the AN/DAN-3

Private defense firms had personnel visiting the project constantly. As Don Duckworth was building the original seeker prototype, a Willys/Overland employee named William F. Osborne was working on infrared homing devices. Just how closely Osborne's work approached McLean's is not known, but it is reasonable to expect that there was little overlap originally. General Tire and Rubber was deeply involved with China Lake; in fact, the company operated the NOTS Pasadena Annex under contract, and it also had worked on the canceled NOTS Air Missile. In 1947, when General Tire and Rubber acquired Willys/Overland's infrared division, Osborne came with it.

Osborne would later claim he had developed an infrared homing device in September 1947, which he had discussed with others at General Tire and Rubber and—according to the testimony of other company personnel—with people at China Lake in December 1947. The China Lake personnel included McLean's department head, A. H. Warner, and L. T. E. Thompson (then technical director), but not McLean himself. General Tire made a formal proposal involving this device to BuOrd in December 1947 or January 1948. In January 1948 Osborne visited Washington to show BuOrd officials drawings of this device. They steered him to McLean, whom he visited presumably before 12 January 1948.[12]

McLean explained the China Lake seeker to Osborne and gave him copies of notes and sketches. McLean later remembered a telephone call from Osborne; McLean thought the call showed that Osborne did not really understand the invention. This lack of understanding may have led him to loan Osborne seeker parts for experiments at General Tire; other accounts say the seeker parts mysteriously disappeared after the visit. Shortly after, General Tire built a larger model. Further refinement and testing led to the company's AN/DAN-3, which by September 1949 was undergoing captive carry flight tests at Point Mugu. These tests apparently went well.[13]

[53]

Did Osborne give McLean the basic idea for the Sidewinder seeker? This seems unlikely. Certainly McLean's manifest ability would make him a better candidate for discoverer than Osborne. Further, McLean thought Osborne had failed to grasp even the basics of the China Lake seeker. One of McLean's supervisors, however, may have mentioned the General Tire project to him, and this may have stimulated his ideas. McLean might have forgotten that he heard about Osborne's invention. Even so, Osborne's idea was probably not unique. Inventions frequently have multiple roots, and it may have been only one of many currents that attracted McLean's attention to a rotating gyro using infrared. It seems more likely that General Tire decided to build on McLean's invention. McLean was much more careful after this experience.[14]

In any event, the AN/DAN-3 seeker developed by General Tire—later Aerojet Corporation—soon began to compete with Sidewinder. While McLean felt that the AN/DAN-3 essentially had been copied from early Sidewinder concepts, as late as October 1951 some groups outside China Lake (such as members of the Research Development Board) still felt the AN/DAN-3 seeker was equal to or better than Sidewinder's.[15] Eventually the navy decided on a conference to consider what ought to be done with the two missiles. McLean and Howard Wilcox were there to defend the Sidewinder concept.

Aerojet was arguing what they could do and what they couldn't, and so forth, and at that time, I [Wilcox] whispered to McLean. I said that I've figured out a way to get around this problem with background clutter that you get against the ground and against clouds and so forth. He said, "Well, what is it?" I drew him a checkerboard reticle. I said, "Should I talk about this in the meeting?" He said, "Sure, why not?" So I did. And the meeting concluded that hey, the Sidewinder's way out in front of the AN/DAN-3 thing, and so they just swept the AN/DAN-3 under the rug at that point, and that was the end of it.[16]

Aerojet would later sue China Lake for patent infringement, but it would lose.

The checkerboard reticle cut down the background clutter significantly. The detector still needed work, however, to discriminate against clouds. (Clouds remain a problem today.) Nichols quickly determined that the background infrared radiation from sun and clouds was very different from a jet exhaust: A jet's tailpipe emitted pure blackbody radiation at 630 degrees centigrade, while the cloud reflection had the same spectral characteristics as the sun. The sun's radiation fell off quickly in the near-infrared region, but the exhaust pipe emitted strongly in this wavelength region. The team desperately needed a filter that would pass only the correct frequencies.

Nichols tried a number of Corning Glass filters but none succeeded. Finally, a chance encounter with a Dr. A. L. Turner at Bausch and Lomb turned up an experimental interference filter that worked well. With a gift of a quarter-sized piece of the new filter, the group flew home and promptly installed it. The filter worked so well that the seeker achieved its first successful tracking of a jet flying against a cloudy background. Nichols was convinced they had turned the corner and their missile would work.

Nichols and Turner experimented further to find the optimum wavelength cutoff for the filter. With the SCR-584 radar mount, they were able to evaluate a filter within minutes and change it as the aircraft was setting up for its next pass. An exhaustive survey suggested that the optimum multilayer filter would cut off wavelengths below 2.05 microns. The dome material finally selected was Corning type 0160, chosen because it was cost-effective and transmitted well in the 2.0 micron to 3.0 micron wavelength range.[17] Hughes Aircraft, developing an infrared version of Falcon, had meanwhile decided on the basis of paper studies that the proper cutoff was at 1.85 microns. The Hughes filter, however, did not discriminate as well.

China Lake's success in developing IR technology so quickly owed much to such simple concepts as the SCR-584 pedestal. According to Nichols:

I attribute all that to the old radar facility that has been the workhorse over the years. You can do this filter selection on paper, at a desk, having the spectrum of the background target, but there is no substitute for going out and doing it against the real thing. That we did! We had everything we needed here at the weapons center. We had airplanes, a place to fly them, the aircraft ranges, and we could talk to the pilots. We could tell them what we were trying to do. We had air controllers out on the ranges that knew what we were trying to do, close cooperation. We could sit and wait for the right day, with these puffy cumulus clouds, and we were all ready to go the minute we had the "bad background" we wanted. . . . It was just this set up that let us study these parts and choose the proper ones.[18]

[55]

Hughes and General Tire had nothing like this. Nichols suspected that the inferior reticle on the infrared version of the Hughes Falcon was attributable to their long cycle of design and testing. "If we had worked like Hughes and Aerojet," Nichols noted, "it would have been like wearing boxing gloves. As it was, we had the whole thing in the palm of our hands. That's why a facility like [China Lake] is so valuable."[19]

The SCR-584 pedestal was indeed a showpiece. Visitors insisted on seeing it at work. A photograph of the seeker showing the image of the airplane on the reticle provided the final touch. If two planes were on the reticle, it became obvious which plane the seeker was tracking.

Late Hours

Working on the heat homing rocket meant working late. McLean seemed to ignore the clock, but others suffered. Nichols dreaded McLean coming in to see him late in the afternoon.

In the early days, when we were getting these first tracking films that we needed so bad, Dr. McLean would show up about 4 P.M. We knew we'd had it, because you knew you wouldn't go home until four in the morning. He'd never run down. I never knew where he got his energy. . . . He'd always have one more thing he wanted to try. Sometimes we'd be out on the ranges crawling around on our hands and knees in the dark, wiring stuff together, the wind screaming. It didn't seem to bother him any. You didn't mind it so bad when he was out there too.[20]

Dave Simmons remembers McLean out on the Baker 4 test track at two in the morning, in shirtsleeves, waving a cigarette to see how the tracker was working. Simmons also noted that the late hours sometimes reflected bad planning. "In the early days of Sidewinder, it really was a circus. They were technically brilliant people, but showed absolute, total disorganization."

McLean's Philosophy of Experimenting

McLean approached problems in two steps, each designed to gather information. The first was a general background, information-gathering phase during which he often operated in a leisurely way, following his own interests. With the Sidewinder, the ideas slowly accumulated until the problem stood out clearly. For McLean, a system had to have "a proper gestation period and a natural birth." During this phase there was a lot of consultation. Tests and peer review were a natural part of this "proper gestation period." He believed it was dangerous to start too fast.[21]

The emergence of a solution triggered phase two, when the focus shifted to the key subsystems on which the solution depended. One had to identify the critical problems, the ones that would hold up the progress of the entire system if they were not solved. These critical areas of ignorance then became the focus for phase two, which operated at a brisker pace.

Once there was a clear idea of the problem, McLean moved rapidly. It was part of the China Lake culture to proceed at breakneck speed once the goal was clear. In this rapid experimentation phase, his primary method of "idea-building" was to construct and test prototypes. Reading the literature was good, but when one reached the point where others had failed, it was time to test.

He insisted on not getting too much data. The universal anxieties felt by all engineers cause most of them to overspecify their data requirements. McLean recognized this, together with its adverse consequences: the time and money wasted gathering all the data. The emphasis was on getting just enough data to see where one should zoom off to next, an approach that depended heavily on McLean's intuition. An R&D group did not have the leisure to develop a complete picture of reality, McLean believed; it had to operate on a partially glimpsed picture.

When Wilcox designed an amplifier that would work but couldn't get enough power gain, for example, McLean came into the laboratory [57] one weekend and solved the problem on his own. Using positive feedback to boost the amplifier's power, he designed a circuit with three turns around the magnetic core, which allowed the output current to flow through these turns to augment the control current. This provided an amplifier that had a gain of three hundred. To Wilcox, this was "black magic." McLean similarly invented the "shorted turn" to eliminate a flat zone in the amplifier's response to the control current.

Wilcox was a high-powered physical analyst; he tried but failed to understand just how the shorted turn worked, in spite of McLean's attempts to explain it to him. Intuition, not formal analysis, had solved the problem. McLean's intuition suggested the shorted turn; it worked; and finally it went into the missile.[22]

This was not an approach for the faint of heart.

McLean's philosophy also avoided imaginary problems. He preferred the simplest approach possible, and when problems with this simplest path arose, he tried to find the most efficient detour around them. McLean refused to let his ego intrude on the design process. "Wire around it," he would say. "Don't solve any problems you don't have to solve." The key was to get a working product, not to struggle with nature.[23]

McLean had a healthy respect for nature. He saw humans as inventors and nature as the judge. He strongly believed in trying things out rather than figuring them out, which was one of the reasons he liked China Lake and one of the reasons he hated formal requirements in advance of development. What was going to work, McLean thought, could seldom be known in advance. Only when laboratories could experiment freely did they arrive at the best solutions to technical problems. Requirements were appropriate when, and only when, a system had already been proved in the laboratory. Only then could one be sure about what to produce.

McLean found experimentation exciting, and his enthusiasm communicated itself to others. It also provided a sometimes not-so-subtle form of pressure to push the development process along. His management style, with much "walking around," reinforced this pressure. When McLean saw people in the corridor, he often asked apparently casual questions. The responses, however, often shaped McLean's opinions about his subordinates. Those without answers did not do well with Bill McLean.

A Bias for Action

McLean thought experiments should explore the genuinely unknown, not simply confirm well-founded hypotheses. "One way to turn McLean off," Don Moore noted, "was to confront him with the prospect of an experiment whose outcome had been calculated." If you knew, there was no point in experimenting. If you did not know, however, mathematics and analysis were no substitute for experiment.[24]

McLean, educated as a physicist and by nature a systems thinker, paid no attention to the traditional boundaries between engineering disciplines. Although he was more than capable of exploiting the mathematics of all fields, he put little trust in their analyses or deductions—and even less if the mathematics were obscured by the intervening machinations of computer programmers, for whom he reserved special contempt. Their programs often left out things that were inconvenient to calculate, and these left-out variables often turned out to have strong effects in practice.[25] According to Wilcox,

> In connection with the development of Sidewinder missile . . . McLean relied on very few calculations, and neither did the rest of us. We roughed out the problem in terms of calculations, trying to figure out how fast the missile had to fly. . . . We spent little time trying to simulate the aerodynamics, every little twitch in the system. . . . Those are the kind of factors which could make or break the missile as an operational system but which are almost impossible to calculate correctly. So Bill would prefer to build the simplest kinds of hardware, and if it gave signs of trouble, he would rapidly try to correct that.[26]

The key to McLean's success was how rapidly he would dive into the prototype phase and thus correct his mistakes. Ordinary R&D work goes step-by-step. McLean, however, took giant steps, landing far ahead of where routine development would have gotten him. The ability to make such intuitive leaps and test them through prototypes was key to the rapid development of Sidewinder. While no critical step was skipped, very few unnecessary steps were taken.

To illuminate the critical problems, McLean often delegated them to small study groups. According to Chuck Smith, [59]

> [in developing the Sidewinder] McLean identified for himself, and later for us, a number of critical areas which had not yet really been looked at in the detail he thought necessary to bring forth a system. One was a torque-balance servo system, which he considered a primary ingredient for the system. Another was a simple power supply which would use a gas generator to drive a turbo alternator. Finally, there was a unique infrared sensor. You can visualize now that a few hobby shops, as we might say, began to arise, one group of fellows attacking the problem of the torque balance servo, another attacking the problem of the power supply, and finally several to attack the sensor/seeker design.[27]

On critical systems, McLean encouraged competition to compensate for the built-in uncertainty of working on novel systems. At least four groups worked on the Sidewinder seeker. This parallel teams approach provided "strategic flexibility," a phrase borrowed from Art Fry, coinventor of the Post-It Note.[28] Strategic flexibility meant the program would not be forced into a corner or become locked too early into a single approach. McLean insisted on keeping an open mind. When Louis Pasteur said that "fortune favors the prepared mind," he meant the mind open to new observations. Openness means willingness to consider alternative courses of action.

A Heat Homing Rocket

With the success of the early studies on infrared, McLean was sure that a heat homing rocket would work. One year after his original sketch, he had developed a set of new ideas on how to get the infrared seeker to work and had recorded these ideas in his notebook. The stimulus for the new ideas was a discussion with Jack Rabinow in Washington. Rabinow suggested an unusual design for a course-keeping gyroscope supported by a central ball instead of gimbals, an idea borrowed from the Bat missile, on which Rabinow had been a consultant.

McLean recorded the suggestion in his notebook on 18 November 1948. The ball gyroscope allowed the project to go forward and would go into what became the Sidewinder 1 missile. The seeker's "free gyro" was a key element. The missile could then simply follow where the tracker pointed, rather than figuring out where it was in space. This was a key move. It saved a huge mass of electronics needed to keep track of where the missile was in space. The Sidewinder in fact never did "know" where it was in space, except in relation to the target.[29]

McLean didn't like ordinary gyro gimbals because they were complicated and expensive; they led to unwanted torques and motions as compared to a single spinning rigid body. McLean liked the central ball gyro and even figured out an ingenious way to manufacture it. Unhappily, the ball gyro caused serious problems, and McLean's enthusiasm was finally checked by the test results. While it was used on all the Sidewinder 1 missiles, a more conventional "Scotch-cross" internal gimbal system later replaced it in the Sidewinder 1A.

On 20 June 1949 McLean formalized these ideas in a sixteen-page proposal for a heat homing rocket that China Lake forwarded to the Bureau of Ordnance. This was the key proposal for the missile that became the Sidewinder. Although some important changes later took place, many of the major features of the future missile were spelled out in the initial document. The missile would have

infrared guidance, using a gyroscopically stabilized and electronically precessed seeker

forward guidance fins (canards) driven by pistons

a hot-gas power supply derived from rapid-burning grains to drive the pistons; battery power for the tubes also came from a gas-grain driven turbine

a servomechanism system producing a torque on the fin, rather than a specific deflection angle (the "torque-balance servo system")[30]

Some ideas would later be dropped. These included: two gyroscopes (one was enough), a separate gas grain for each piston (a better system was a big single grain for all four pistons), and small, prism-shaped canards (replaced by larger, fin-shaped ones). Many other features would be changed, enhanced, or added before the rocket reached its final form.

McLean had submitted these plans none too soon. General Tire and Eastman Kodak were closing in.[31]

Fox Sugar 567

The heat homing rocket was starting to take on a more definite shape and now needed a name. Ted Whitney had suggested "Flamingo" as a contraction of "Flame and Go," and other names were proposed. Another member of the Aviation Ordnance Department, Gil Plain, had noted during a division meeting in about 1948 that the Sidewinder rattlesnake (the horned rattlesnake, *Crotalus cerastes*) detects the infrared radiation emitted by its prey. "Sidewinder" seemed appropriate as an expression of China Lake's identity and desert location, so the "heat homing rocket" became the Sidewinder project on 27 November 1950.[32]

In 1951, however, word "came down through the navy that there was no such thing as a Sidewinder missile." China Lake was told to lie low, cease talking about a missile, and speak instead of "feasibility studies." The reason, apparently, was that the Truman administration wanted to cut a large amount from the budget for Sidewinder, amounting to cancellation of the program.[33] So Sidewinder dropped off the budgeteers' radar scopes. Previously, it had been both Local Project 602 and Feasibility Study 567, each an earlier stage of the project. (The numbers appear to have originated from the telephone extensions of the experimental officer at China Lake, 71567 and 71602.[34]) For about two years, the Sidewinder project was known as "Fox Sugar 567." By that time there was no need to hide it and no way it could be hidden, so Sidewinder emerged again.

The Weapons Shop

Behind a black wall of secrecy, the U.S. is climbing slowly
toward a new level of warfare. In every U.S. factory, every
technical institute, and every electronics laboratory, the military
phrase of the day is "guided missiles."
Jonathan Norton, Newsweek, 21 May 1951

The "Intelligent" Fuze

China Lake continued work on seekers for its outlaw missile. This work
was funded in part by money for fuzes—after all the missile was "a
fuze with an inherent tendency to reduce dispersion."[1] While McLean
had spelled out the basic idea for a seeker in his "heat homing rocket"
memorandum, he had several teams pursuing different approaches to
keep his options open.

As results came in from Nichols's infrared "target survey," McLean
decided to speed things up. A revealing 26 May 1950 memorandum
from McLean concerning project requirements is peppered with comments from L. T. E. Thompson, the technical director of China Lake.
The comments remind McLean that this is an *exploratory* program.
Thompson wanted to make sure McLean did not get overcommitted
to a dubious project.[2]

McLean had no such doubts. He was impatient to get going on
development, because he was certain Nichols's research would show
that infrared detection worked. Later memorandums show that McLean

already was well into developing the components for his heat homing rocket. Even the first of these (26 September 1950) shows a long list of design objectives for each of the missile's subsystems. Three different designs for seeker heads were being explored simultaneously— two by China Lake and a third with the help of Avion Corporation. The second memo to Thompson, dated 25 October 1950, provides a snapshot of a team in the process of formation:

> The organization required to provide the subject control and fuze will require the assistance and active support of nearly all the departments on the Station. It is the purpose of this memorandum to submit for your approval a breakdown of the various tasks to be accomplished into integrated units. The personnel required for these tasks are also indicated. It is my intention to act personally as project engineer to co-ordinate the activities of the various groups and to make final decisions regarding incorporation of experimental designs and results into the prototype model.[3]

The memorandum indicated that three seeker heads for the rocket would be developed:

The A seeker and amplifier were being developed in conjunction with Avion Corporation in Paramus, New Jersey. McLean favored this design (presumably A for "Avion").

The B seeker and amplifier project at China Lake, directed by Roger Estey, used a stationary armature to spin the gyro, external gimbals, and magnetic precession. The B head, usually attributed to Luke Biberman (thus B), actually was used on the first guided missiles.

The C seeker and amplifier head project at China Lake, directed by Jesse Watson, used a central spherical bearing.

Meanwhile, Aerojet also was working on a D head (no doubt the basis for the AN/DAN-3), and Eastman Kodak was developing an E head.

The 25 October memo also discusses how the team would investigate the OMAR air-to-ground missile, with a beam-riding guidance system, under Walter LaBerge. Other teams, involving twenty-one other personnel, were to work on propellants, production and engineering design, aerodynamics, and target survey. Some twenty-four engineers and technicians were involved in the project at China Lake at the time, with more at Avion. As weapons systems go, this was a small group—but it had energy and direction. Much of the team's success can be attributed to China Lake itself. It was a laboratory where people were encouraged to find work that suited them and do the things that needed doing.

Wilcox and LaBerge Arrive

By the end of 1950 all the key parts of the weapon system were under development: seekers, canards, rollerons, generator. But McLean needed more high-level talent to pull the parts together into a missile. At this point, he was fortunate in recruiting two able assistants—Howie Wilcox and Walt LaBerge—both Ph.D. physicists.

Wilcox was short, energetic and possessed unlimited self-confidence not only in his physics but also in his prowess at games; he would play tennis into his seventies. He had been trained as a physicist at the University of Minnesota, where he graduated magna cum laude in 1943. After a year of graduate education at Harvard, he spent the rest of World War II at Los Alamos, New Mexico, as part of the Manhattan Project working on the atomic bomb.

After the war, he finished his graduate education at the University of Chicago, where he got M.S. and Ph.D. degrees in nuclear physics, studying with Enrico Fermi. In 1948 he became an instructor in the Physics Department at the University of California, Berkeley; two years later, he was made an assistant professor. This was the McCarthy era, however, and in the controversy over loyalty oaths, Wilcox soon found himself in an untenable position. He resigned in protest.

Despite his considerable nuclear experience, Wilcox decided the field was too crowded. Guided missiles looked like the coming thing. A year later, *Newsweek* would note: "The [guided missiles] program has already drained the country of specially qualified scientists. Every missile plant and laboratory has a welcome for the dewiest young technician." So he approached both Convair and Hughes Aircraft for jobs, but his interviews there failed to excite him. Then his friend Will Stark got him an interview at China Lake.[4]

Arriving for the interview, Wilcox found the desert base reminded him of Los Alamos. He talked to many of McLean's associates and just before noon met McLean himself. Driving along with McLean in his dirty blue 1939 Ford coupe, he had to concentrate on the conversation to stay up with it. At lunch, Wilcox learned that all the men he had been talking to that morning worked for McLean, who was not only the division head but was also leading the guided missile project himself. He learned one other thing. Just before lunch in the cafeteria McLean "appeared to lose consciousness. But the waitress merely murmured something about this sort of thing happening frequently, put a couple of heaping teaspoons of granulated sugar in McLean's cup of coffee, and then pulled his head back to urge some of the sweetened coffee between his lax lips. Within a few minutes he began to recover." McLean explained that he was a diabetic and sometimes went into insulin shock.[5]

Back at Michelson Laboratory, McLean explained the Sidewinder project in more detail. McLean posed some problems to test Wilcox's acuity, and Wilcox came up with useful suggestions. Finally, Wilcox learned that only two dozen people were working on Sidewinder, thus it seemed to offer an excellent opportunity for a bright young physicist. When China Lake offered him a job—for much less money than Convair or Hughes—he took it.

Wilcox became the project's number one analyst, highly respected for his mathematical abilities and his ability to communicate with McLean, with whom he shared a strong mental and emotional chemistry. Wilcox sometimes acted as interpreter for McLean to those without the ability to understand the sometimes obscure suggestions McLean would make. Wilcox thrived on the mental table tennis he played with McLean. He described how, when he entered a room with McLean, he could feel a kind of radar playing over him. McLean wanted to know what Wilcox thought, what new ideas he had come up with. McLean himself often bubbled over with new ideas on returning to China Lake from Washington. Bob Blaise once suggested to Hack Wilson, associate technical director, that they should always tape-record the first five minutes after McLean got off a plane, to catch all the ideas he came back with.

Walt LaBerge, tall, slim, and handsome, was more soft-spoken and slightly younger than Wilcox, and had a Ph.D. in physics from Notre Dame. LaBerge, like Wilcox, possessed administrative talent. LaBerge's immediate responsibility was to develop the OMAR beam-riding, air-to-ground missile. After several months of study, it was clear to LaBerge and Wilcox that OMAR would not work. So it was canceled, and LaBerge transferred to the Sidewinder project. When the team suggested canceling the OMAR project, Washington resisted, but McLean was fond of quoting Charles "Boss" Kettering that "the most important thing in research is the proper burial of a dead horse."

On Sidewinder, LaBerge saw his role as finding out who really wanted to do something and to get him doing it. He was also a superb technical salesman. When LaBerge joined the project, he and Wilcox shared duties, but they soon agreed that Wilcox should be the project engineer and LaBerge the missile engineer, responsible for all hardware and interfacing with the contractors such as Eastman Kodak and Avion.[6]

McLean was fortunate to get these two men. After their work at China Lake, both LaBerge and Wilcox would hold key positions in the Defense Department and in private industry.

"The Gadgeteers Were in Full Swing"

Like Tom Sawyer, McLean knew how to build a team. He would start an interesting project, then tempt others to join him. He said little, but his vision was compelling. China Lake was a "matrix" organization in which project teams like Sidewinder were expected to borrow expertise from departments such as research, engineering, and test, which constituted the basic hierarchical structure of the base. The departments controlled pay and promotion, but projects, both on and off the base, drew from them. The projects were temporary networks of personnel that shifted over time, while departments were permanent structures.

A matrix organization inevitably generates tension between its projects and its disciplinary departments. Functional departments deliver expertise; project groups deliver prototypes. Each is likely to see its own role as primary. Two reporting systems create two sets of loyalties. Strains arise from these dual loyalties; projects and departments struggle over funds and prerogatives. Who gets credit when a project succeeds also is a problem. Functional work yields *contributions* to systems (often referred to at China Lake as "cats and dogs") that later may be hard to identify; whereas a *project* represents a free-standing system whose success will clearly reflect on the team creating it. This difference in visibility guarantees jealousy over credit given. Matrix organizations are inherently unstable systems that require the utmost in judgment from top management.[7]

As the project proceeded, McLean tried to find qualified people from the functional departments to assist him. McLean often got these people to join and assist the project, sometimes more intensely than their original brief intended. Ted Toporeck, head of the Test Department, sent engineer Bob Blaise over to work with McLean; Toporeck thought the Sidewinder project would not last long. According to Blaise:

[67]

> One day early in 1952 the Head of the Test Department called me to his office and closed the door. He said, "Bob, the Head of the Aviation Ordnance Department, Bill McLean, has a so-called infra-red

Key personnel of Sidewinder project. *Left to right:* Edwin G. Swann Jr., Lee Jagiello, William B. McLean, Howard A. Wilcox, Walter LaBerge, Jack Christman. Courtesy of Ernest Cozzens

homing air-to-air missile that he is putting together. He needs someone from our department to handle the testing for him. I want you to
take a little time off from the Terrier program to do his testing for
him. It shouldn't take long, then you can resume the work on Terrier." That was the last time I fired a Terrier and the beginning of the
most rewarding experience of my life.[8]

Blaise never went back. Events like this did not endear McLean to
the functional department heads. As the head of the Ballistics Division put it, "The gadgeteers were in full swing and this was all that
mattered."[9]

These conflicts could and did become intense, but McLean usually
won them. McLean believed that projects should determine the organization, not the reverse.[10] If a key person was needed, McLean tried
to recruit that person. He tried to create whatever structure was necessary to build ideas. Functional departments resisted these personnel forays fiercely and fought back as they could.

The Night Factory

They worked hard. Howie Wilcox describes going back to the laboratory in the evening after dinner. In the parking lot would be a little
knot of cars: the Sidewinder team. The ever-present background of the
Korean War and the emergence of the MiGs provided a constant
reminder of the need for air superiority. Requests for assistance from
naval units operating off Korea arrived regularly at China Lake, another
reminder that engineering had serious consequences for men at sea.
Although American pilots did well with guns against the MiGs in
Korea, air-to-air missiles were obviously the next step. The arrival of
the jet age meant that pilots would require something more than cannon and unguided rockets. The Sidewinder team was trying to provide that something.[11]

Critical projects focus these intense energies. Chuck Smith, a young
engineer at the time, later commented:

There was a feeling, not that you worked at a big naval test station,
but that you were part of a close-knit group, a university-like atmosphere. . . . Communication between the higher levels and the lower

levels of the community was very good. . . . If you thought it was a problem that [McLean] might know something about, you could go over and talk to him about it. And he would show great interest in what you were doing. So the result was that each guy working on that program developed a real commitment to get his part of that job done. And whether it took you eight hours a day or whether it took sixteen hours a day to get it done, you worked until you got your part of it done—so your part didn't lag anybody else's part.[12]

[69]

Sometimes the work took the group out onto the rocket ranges at night, searching the sand for fragments of a crashed missile. Scientists and technicians would search on their hands and knees. For the technicians, this was overtime work with no overtime pay. McLean often led these searches, and Rod McClung once commented that "if it wasn't for Bill McLean, I would tell them to shove this thing."[13]

Getting Contractors Involved

Once serious development of Sidewinder began, McLean sought assistance from other laboratories and outside contractors. In some cases, several potential sources were encouraged to offer design ideas; sometimes an internal team competed with an external group. The net result was a process that had, in the beginning at least, many options. "Different approaches to the lead sulphide cell, the rotating chopper, the servo valves, the amplifier, the generator, and the propellant grain were, at some time in the course of development, pursued by groups at [China Lake]. The major set of parallel efforts was directed at the seeker head itself . . . ; two contractors and two groups at NOTS explored competing paths."[14]

Philco and Avion were the two major contractors for the seeker, although only Avion did initial development. But other contractors were working on the lead sulphide cell (Eastman Kodak), the fuzes (Eastman Kodak again, Bulova Laboratories), the optical filters, and so on. Other laboratories, such as the Naval Ordnance Laboratory at Corona, California, also were working on system components. While China Lake did much of the development, in some cases the China Lake team were the inventors and the other groups the developers or mechanical–electronic designers. The Sidewinder team would think up a gizmo, and the gizmo would be perfected elsewhere. The team could concentrate on overall direction of the development effort, but they became more dependent on outside groups.

Unlike many of the others, the Avion Instrument Company was a very small operation in Paramus, New Jersey, with about twenty-five people, including nine engineers. Avion's director, Richard F. Wehrlin, who had been chief engineer on the development of the Norden bombsight of World War II fame, had heard about McLean's missile plans and knew that China Lake was looking for an outside contractor for the seeker. Don Friedman, a young electrical engineer, wrote Avion's proposal, and China Lake in 1950 gave the company twenty-five thousand dollars (a large sum at the time) to complete a six-month development effort on the A head.[15]

Friedman's group quickly decided that an air-driven design would not work well. Instead, it developed a ball-gyro system in which the gyro formed a closed magnetic circuit with an external stator so that it could be driven by electrical forces. Four coils placed ninety degrees apart around the gyro and alternately turned on and off would get the gyro (a magnet) to spin. This made the gyro essentially a motor with its only fixed part the point where the ball was attached. A fifth coil would precess the gyro; no target, no precessional forces, and the gyro would spin without any preferred orientation. With a target, however, the exterior core became a skewed electromagnet, thus exerting a precessional force on the gyro, causing it to tip toward the target. The design called for using the exterior core as part of the magnet's field, but metal shapes near the gyro produced eddy currents that seriously interfered with the gyro's operation. With six weeks to go before Avion had to present a design, Friedman was in trouble. Then he had a "ten-minute thought" as to how the problem might be solved.[16]

Friedman whipped up a prototype, using epoxy as the container rather than the eddy-current inducing metals. The four motor coils were embedded in the epoxy at ninety-degree angles, and the fifth (precessing) coil spiraled around the other four. A sixth, sensor, coil was added, which would not have any current except that induced by the magnet. The sixth coil's function was to sense the orientation of the magnet and thus to "know" what commands should be given to the magnet to get it to precess.

McLean arrived at Avion on schedule, and the prototype was ready. The system would track a cigarette or a light bulb, but sensitivity was not good and the system was unstable. As part of an aircraft-carried missile, the magnet-gyro would have to be surrounded not by epoxy but by metal. To Friedman, the prototype was thus interesting but ultimately useless. McLean, however, was unconcerned. "If it's flying in space, there won't be any metal around it, and there is no need to precess the gyro until after the missile is fired." The key to allowing the system to work, then, was to keep the gyro pointing straight ahead—caged—until the missile had been fired. The missile would lock on the target when the pilot pointed the plane at it. Friedman had not realized that McLean was willing to let the pilot find the target and then point the plane.

[71]

Friedman was elated that McLean had faith in him. "I was just a kid," he remembers. Even more important, McLean gave Avion another twenty-five thousand dollars to refine the design. Friedman rapidly gained the team's respect and worked closely with its members.

Contracting out was not always so successful. The missile's launcher, assigned to the Ballistics Department, did not work, so Lee Jagiello and Wilcox took up the design, which Wilcox remembered finishing in three months. The Sidewinder team always thought that if outsiders failed, they could take the design away and do it themselves. This attitude did not endear it to other groups.[17]

As work proceeded, a network of supporters evolved. More people and more money were now in the picture. The missile was becoming more real but also more of a threat to other programs. Previously, Sidewinder had seemed so insubstantial that it could be ignored. That had changed; virtually all departments at China Lake now seemed to be working on Sidewinder.

A Visit from the Missile Czar

In 1951 Washington suddenly took note of what was going on at China Lake, and K. T. Keller, formerly head of the Chrysler Corporation and now the Pentagon's new director of guided missiles, headed for the high desert. Keller knew that shortages of critical components meant the country could support only so many guided missile programs, so he began to evaluate existing programs.[18]

No doubt Keller was selected because he was seen as action oriented. When Wilcox met him in early 1952, however, the missile czar struck him as "a crusty, foul-mouthed, self-confident, but not very bright man in his late sixties . . . possessed of a truly monumental potbelly." A story was circulating that Keller had casually destroyed the prototype of the vitreous radome for the Hughes Falcon missile by dropping it on the floor, saying that "this damn thing will never stand handling," while Hughes personnel watched in horror.[19]

McLean was in the hospital at Santa Barbara for his diabetic problems when Luke Biberman got the bad news that Keller was coming to China Lake because they were "building missiles and spending navy money for pieces of missiles, and so he was coming to shut the place down." At the time, Biberman was developing the B head, which was further along than the A head favored by McLean. Biberman remembers:

> I didn't go home. I had two and a half days to build one of these things. Ed Swann and Johnny Miata and I put in some pretty long hours. I guess that they had just gone home to get some sleep, and I was determined I was going to have this model working before this guy showed up. I had just soldered the last two connections on the phase discriminator . . . and I had never turned it on. I looked up to see someone walking in the door with an escort. This was the missile czar being sent by Charlie Wilson [secretary of defense] to examine what we were doing.
>
> I looked at the guy, and I was really intent on what I was doing, and he was smoking a cigarette. And I said, "May I have that please," and he thought I didn't like the smoke. He gave me the cigarette, because he couldn't see an ashtray. I cranked on the power, and I opened up the gas to run the servo and I began to move the cigarette in a circle. And the gyro moved in a circle. And the fins moved up and down and sideways, as this thing went around in a circle. And he didn't notice—although I did—that the fins were going down when they should have been going up, and they were going left when they should have been going right. They were exactly 180 degrees out of phase, which meant I had soldered the wires of the phase discriminator backward.

The ex-Czar of Guided Missiles visits NOTS

"It's getting more complicated, isn't it."

K.T. Keller
1952

[73]

K. T. Keller, the nation's "missile czar," was not always a fan of Sidewinder. Cartoon by Howard A. Wilcox

Nevertheless, he was watching the cigarette. He said, "Make it go up and down, and make it go left and right." Only up and down fins worked, or only the right and left fins worked, or they all worked together in a circle. His next question was "How much money do you need to make it fly?" About that time—I tell you, there are some wild moments in this business—in comes Bill McLean looking lousy, but he had got the news that this guy was here. And he came in and when he saw the fins going up when they ought to have gone down, he just about died. He walked over to me, and he shut off the nitrogen; he shut off the power. He looked around and said, "Where's the phase discriminator?" Bill switched the wires but [Keller] didn't need to see any more. He and Bill walked off together and talked about how we were going to proceed from here.[20]

This was only one of Sidewinder's many close calls. And Keller, satisfied for the moment, would later change his mind.

[7]

Systems Engineering

Beauty is momentary in the mind,
the fitful tracing of a portal,
but in the flesh it is immortal.

Wallace Stevens

Systems Engineering for a Heat Homing Rocket

Sidewinder had started with a clean sheet of paper because McLean knew that too much borrowed technology can compromise a design. But he decided early to build the missile around a standard propulsion system: the five-inch high-performance air-to-ground (HPAG) rocket motor, a system China Lake had developed and was refining. China Lake personnel were familiar with the rocket and its aircraft launch equipment was readily available.

The HPAG fuselage was large enough to house a powerful warhead. Just as important, it had a long-burning motor with relatively slow acceleration, which meant less shock in launching for the Sidewinder gyroscope and other delicate components.[1] The standard airframe, however, had to accept many nonstandard components: the seeker head, the control surfaces, and the torque-balance servo. Radically new weapons often require a whole series of special inventions and the Sidewinder was no exception. Although McLean often described the Sidewinder as just a rocket with a guidance system, in fact the missile was a remarkably well integrated system. McLean believed that every part ought to serve more than one function. In Sidewinder, Henry Swift observed:

> The pistons that moved the aerodynamic controls were also the solenoid that actuated the valves on the pistons that controlled the gas pressure on the pistons. . . . The use of high-pressure gas from a piece of burning rocket propellant–like material spun a gas turbine at 60,000 rpm to generate the electrical power needed, and also provided the motive power to operate the canards. The cylinders containing the pistons doubled, probably, as structural elements, etc. The infrared radiation collection mirror was also part of the spinning gyro, as was a permanent magnet that was also the rotor of the "motor" that spun the gyro. The magnet was also the element by which forces were applied electromagnetically, to turn the mirror/telescope, so as to track the target.[2]

[75]

So Sidewinder was neither a lucky combination of off-the-shelf items, nor—as Bob Blaise sometimes said—did it involve divine intervention. The truth is that McLean was an exceedingly gifted systems engineer.[3] The overall missile design would survive virtually all the electronics revolutions of the late twentieth century, although subsystems have improved with each successive model.

Sidewinder set the standard for air-to-air guided missiles. The nose-mounted seeker tracked the target and generated signals to the control surfaces that steered the missile onto a collision course. Aft of the seeker was the control unit—an elegant servomechanism that moved the steering canards. Next came an electrical turbine, powered by a burning gas grain, that generated the electricity to move the canards. Gas from the grain also drove the pistons in the servomechanism.

Behind the generator were the contact and proximity fuze systems and the warhead. The aft-most portion of the rocket casing included the rocket motor (also a solid gas grain) and the fins with their rollerons. The fins provided directional stability, and the rollerons slowed the missile roll rate.

A Tracking Head

Several seekers were considered, but I concentrate on the one eventually used—which McLean himself largely designed: the A head. It

PHYSICAL CHARACTERISTICS

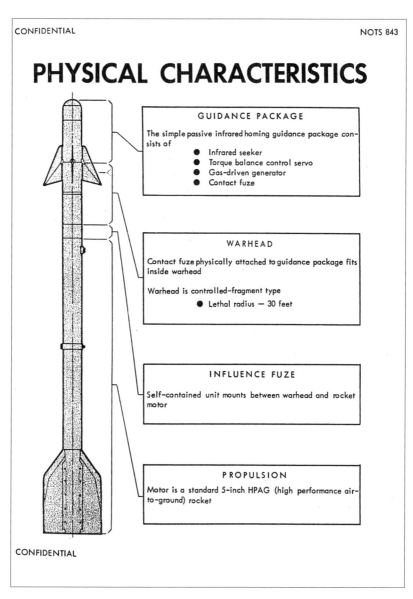

GUIDANCE PACKAGE

The simple passive infrared homing guidance package consists of

- Infrared seeker
- Torque balance control servo
- Gas-driven generator
- Contact fuze

WARHEAD

Contact fuze physically attached to guidance package fits inside warhead

Warhead is controlled-fragment type

- Lethal radius — 30 feet

INFLUENCE FUZE

Self-contained unit mounts between warhead and rocket motor

PROPULSION

Motor is a standard 5-inch HPAG (high performance air-to-ground) rocket

Sidewinder components diagram. Courtesy of U.S. Navy

Electronics module for the Sidewinder 1 (AIM-9A). Courtesy of Frederick Davis

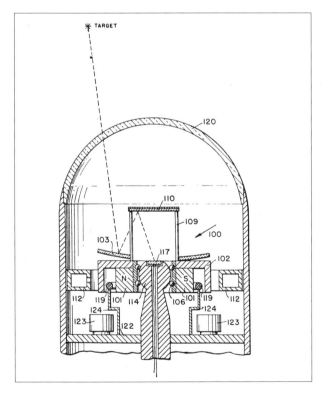

Sidewinder seeker diagram from early patent drawings. Courtesy of Howard A. Wilcox

was a remarkable device: It detected an infrared signal, pointed itself toward the signal, then got the missile to turn onto a collision course with the signal.

The key was the lead sulfide photocell, whose resistance changed when struck by infrared light. When the target appeared in the system's optics, the current across the cell surged. The optical system, a Cassegrain telescope, used a primary mirror and a secondary mirror to concentrate the light. The mirrors concentrated the image of the target on the photocell.

Once the seeker detected the target, it had to decide target location in relation to the missile. Initially, the pilot put the aircraft's nose (hence the missile's nose) on the target, essentially pointing the optical system. After launch, however, the optical assembly had to stay on the target regardless of missile motion. To achieve this, the entire optical assembly became, in effect, the rotor of an electric motor. As the rotor assembly spun seventy times per second, it was directed by precessional torques and thus became a gyroscope that followed the target.

Tweaked just the right amount electromagnetically, precession forces on the gyro moved it to point toward the target's location. With proper phasing, the new spin axis would point constantly toward the source of radiation. Thus Sidewinder kept its infrared-sensitive "eye" on the target. Initially, one gyroscope was to be used to spin-stabilize the detector assembly and another to keep track of the missile's position in space. Hughes engineers, however, suggested to McLean that a second gyroscope was unnecessary because the missile didn't need to know where it was—it just had to know where the target was—and so McLean eliminated the second gyroscope.

As the target moved and the detector noted the deviation, it transmitted signals proportional to the precession rate of the seeker gyro and in the correct direction; these were used to change the position of the canards and steer the missile toward a collision course with the target.

Note that it was a collision course. The missile didn't point at the target, because the target was moving. Instead, the seeker directed the missile where the target was going to be at intercept, using a technique called proportional navigation. Before Sidewinder was a success, there was fierce discussion about whether proportional navigation would work. Some experts suggested it would cause the missile to overshoot the target. In the end, this turned out to be a nonproblem—but there were plenty of real ones.

Using optics designed by Roger Estey of the NOTS Research Department, Avion Corporation's A guidance head proved to be a sensitive detector. Mounted on a tripod, the head used feedback to track tennis balls, birds—and airplanes. One of the more eye-catching demonstrations—tracking a lit cigarette—fascinated visitors and soon became a legend. [79]

The Avion seeker was only one of several approaches that McLean encouraged. There was Luke Biberman's B seeker and Jesse Watson's C seeker. The B head, with large external magnets, weighed about 30 pounds; the Avion head, by contrast, weighed about 10 pounds. (The whole missile eventually would weigh only 165 pounds.) The C head was entirely mechanical, with so many levers and cams that Nichols described it as "looking like the bottom of a Marchant calculator." McLean preferred the A head, but he kept the other approaches going because he was not sure the A head would work. In fact, the B head's tracking performance was superior in ground tests, but when used on a missile it did not work so well. Designing a seeker head proved to be difficult, but McLean's real effort focused on designing the servomechanism.

Learning to Turn

Getting the Sidewinder to turn properly was a serious problem, whose solution rested with Lee Jagiello, the team's gifted aerodynamicist who came from the Ballistics Division of the Research Department. Wilcox's first encounter with Jagiello was classic. Wilcox, seeking some aerodynamic data, was having problems. "The man working on the project . . . couldn't tell me how accurate the numbers were and he couldn't stand behind them. The guy at the next desk turned around and said they were accurate to 5 percent. It turned out to be Lee [Jagiello]. I instantly shifted my attention to Lee, [who] displayed a very fine intuitive feel for the problem."[4]

Jagiello could answer the question because he had gathered the data himself. From that moment the informal relationship between Wilcox and Jagiello developed swiftly. While his help provided an important boost, things did not always proceed smoothly. One of Jagiello's assignments was to work on canard design.[5]

The fins used to turn the rocket were located near the nose (hence the term canard), instead of near the tail, as they were on some other air-to-air missiles such as the Falcon. The reason for the location near the nose was that Sidewinder had to be disassembled for stowage. Ammunition hoists on the aircraft carriers of the day could handle rounds no longer than 76 inches, but Sidewinder was going to be 108 inches long, which meant the missile could not be stored as an "all-up" round.[6] It was going to have to be disassembled for storage, and McLean was concerned that this would result in plugging and unplugging the electrical connectors between the guidance unit and the fins. He knew this was a common source of malfunctions: the female part of the connector got looser with each attachment. Making the guidance and control unit a single assembly with the canards would eliminate the problem and improve reliability.

How Sidewinder's design was shaped by ammunition hoist dimensions illustrates the pervasive influence of existing technologies on a new system. But also it shows McLean's elegant engineering. As with the five-inch rocket motor, McLean pragmatically elected to fit his new conception into an existing system—and simultaneously improve reliability. There are many examples of a design constrained by existing systems—and thus influenced by the politics of acceptance. The fewer changes that sponsors have to make in their existing equipment, the easier it is for them to accept new technology. When designers really do start with a blank sheet of paper, the users may violently resist. With a hoist system able to handle a larger round, Sidewinder might have been designed differently. When carrier hoists changed, however, Sidewinder stayed the same.[7]

The canards had to be rugged and sophisticated since they functioned as part of a second innovation: the torque-balance servo system that provided the same turning rate *regardless of altitude*. The servo was a key part of the missile; even though LaBerge was formally in charge of developing the servo, McLean decided to design it himself.

The canards adjusted to fit the altitude. In the dense lower atmosphere, a 5-degree deflection would produce a dramatic effect, while

Lee Jagiello, aerodynamicist for the Sidewinder missile. Courtesy of Lee Jagiello

the same effect in the rarefied air at high altitude required a larger deflection. "If, for instance, the actuator is commanded to produce 750-inch-pounds of torque on the crank arm, the fin can come to equilibrium at an angle of about 3 degrees with a Mach 3.0 airstream at sea level. However, at 35,000 feet altitude the Mach 3.0 airstream will not balance the same 750 in-lb until the fin reaches an angle of 14 degrees."[8]

In other missiles, such as the Falcon, the pilot solved this problem by adjusting the altitude control before launch. Another possibility was to put an altitude sensor in the missile. In the radio tube age, however, such sensors would have been bulky, expensive, and unreliable.

Sidewinder did not have to endure such indignities. The torque-balance servo system did not produce a set canard angle, but rather a torque on the fin. Inside the missile, pressure in a cylinder produced a force on a piston, which exerted pressure on a knife-edge to turn the canard. How much the piston turned the canard was determined by the relationship of the pressure inside and the pressure outside: the thinner the air, the less the pressure of the airstream. Thus a given piston pressure produced a bigger deflection at higher altitude. When the airstream's pressure on the fin outside matched the pressure on the piston inside, deflection stopped. How far the canard moved into the airstream, then, was determined by balancing the internal and external pressures on the canard. Just as the seeker made irrelevant the missile's position in space, the control system eliminated altitude from the equation. It was an elegant design.

Canard design was subtle. The lift forces they produced when deflected had to cause equal torques about their hinge lines regardless of speed or altitude. As Wilcox put it, "Lee had to design a control-fin configuration for which the center of aerodynamic lift would be a stationary distance behind the surface's hinge line during all the flight conditions of the missile."[9]

Jagiello's solid grasp of aerodynamics allowed him to solve this tricky problem, with some assistance from the laws of nature. Once seen, the system seems not only remarkably simple but also obvious. It certainly was not obvious to most missile designers, however, who made very large compromises in their own missile systems. McLean recognized the great importance of the design of the servomechanism involved in torque-balance control and so concentrated his own design attention on it. He confessed to Wilcox some chagrin at the attention given to the seeker, when in his mind the torque-balance control system was an equal, if not more impressive, design accomplishment.

Sidewinder Gets a Tail

Tail fins on an airborne projectile keep it pointing straight ahead. Unhappily, there is usually very little they can do to keep the missile from rolling on its axis. In principle there is nothing wrong with rolling. Bullets, as a familiar example, are spin-stabilized—essentially becoming gyroscopes. Interestingly, in the heat homing rocket proposal, McLean had proposed letting the missile rotate freely inside a tail ring with fins on it. This proved impractical because such a missile could not be launched from a rail launcher. More important, it was hard to get the canards to work properly when the missile was rolling. So another of Sidewinder's key design features was a simple device to slow down the roll: small gyroscope wheels—rollerons—attached to its rear fins. This simple device has had a long history in the air-to-air missile field, and rightly so. Missiles moving supersonically, as Sidewinder does during the boost phrase, are likely to have serious stability problems.

The idea for the rollerons came from technician Sid Crockett, who first sketched his idea in mid-October 1949. A wheel, notched around its circumference, sat on a hinged tab on the tail fin and was normally held in place by the air stream. The same air stream spun the wheel and turned it into a small gyroscope. If the missile started to roll, the gyroscope's precession would produce a torque on the tab, which would then flip out into the airstream, opposing the missile's roll; when the missile stopped rolling, the gyro would stop exerting a lateral force and the pressure from the air stream would pull the tab back in line with the fin.

[83]

Although Crockett's original rollerons were crude, the basic idea was sound. Jagiello, who had far more theoretical knowledge, began to investigate the rollerons in December 1949 and eventually got a good mechanical design.[10]

The rollerons, like the torque-balance control system, saved hardware. Missiles without rollerons used complicated internal electronics to prevent roll, all of which consumed space and decreased reliability. In June and July 1950 rolleron tests on missiles without seekers showed that the rollerons did indeed slow the missile's rolling. On 15 July, for example, a test firing of four missile dummies yielded roll rates of less than sixty degrees per second. This was fine because of Sidewinder's limited flight time.[11]

The rollerons were valuable sales tools. Military visitors, who sometimes questioned whether proportional guidance would ever work, saw the rollerons and were impressed. People have instinctive reactions to novel systems. If they can see something that makes sense, they assume what they can't see must make sense, too. The rollerons made sense.

The Washing Machine Wobble Damper

The first working model of the A head arrived in 1951. The seeker functioned well as long as the return from the target was not too far above the noise level. As signals got stronger, however, the seeker would begin gyrating wildly and lose the target. Analysis gave few clues, until one day Fred Davis realized that the oscillations might be caused by a particular gyro frequency.

Wilcox, LaBerge, and Davis immediately went to the laboratory and used a signal source to start the gyro wobbling. The problem turned out to be a frequency that was the difference between the spin frequency and the nutation frequency. As the return from the target got strong, this frequency would get into the system and start the wobble, a serious problem. Large gyroscopes on ships can be damped using a variety of means, since weight and size are not major problems. On a small system like a missile, however, such damping is difficult. McLean and Wilcox were concerned.[12]

They tried several approaches before Donald Steward, a bright technician, came in one morning with a solution. He had spent the night before repairing his washing machine and noticed that his machine, an early model, was damped by three large, heavy balls in a circular raceway; when the load shifted, these balls moved to damp the oscillations. This approach was unlikely to work on a guided missile and Wilcox suggested to Steward that he try a viscous liquid. At first, Steward used a single raceway around the head with mercury in it. When this proved unsatisfactory, a second raceway was added. Then it was suggested that mercury would freeze at the high altitude where missiles might be fired, and Steward wanted to add thallium to lower the freezing point. Thallium, however, was highly poisonous, and China Lake's industrial hygienist objected. In the end, however, they used thallium and the damper worked well.[13]

Steward tried a number of designs and used the one that minimized the wobble. Earl Donaldson remembers that testing the nutation damper in the lab was simple. It would be flipped like a coin. If it successfully flipped, it was not working; if it came down spinning on edge, it worked. There was no theory for the way the mercury damper worked—it was obviously a "de-tuned" element in a tuned system—and even Wilcox, the group's analyst, never felt he really understood how the damper worked.

[84] ## "Independent of Aerodynamics"

While several groups struggled with the seeker, Jagiello was trying to perfect the airframe and its unusual canard configuration. Jagiello was driven nearly to distraction by McLean's idea that Sidewinder should be "independent of aerodynamics." Indeed, McLean told Leroy Riggs (later technical director at China Lake), "with the right servo you could fly a barn door." To an aerodynamicist, those were fighting words;

worse, they betrayed a serious break with reality. Jagiello was concerned that the missile would have too much drag to achieve the required range. While most team members had absolute confidence in McLean, Jagiello fretted. "McLean would come up with some crazy ideas," he recalled.[14]

One of these was the rectangular-planform canards and fins on the early EX-0 airframe. McLean always tried simple solutions first; only if they failed did he opt for more sophisticated approaches. McLean thought that square fins would work and be easy to produce. Jagiello was convinced they would cause oversteer and too much drag; he wanted fins that were more tapered, and finally, after the square fins flopped, he got them.

[85]

Lack of funds meant improvised equipment. One attempt to find the optimum angle-of-attack for the canards involved mounting a half-scale missile on a truck and driving it across a dry lake bed near the base.

Similarly, the "Sunseeker" idea involved getting missiles to track the sun. Since the team could not afford any target drones, McLean thought he would use the sun as a target. Firing a round upward toward the sun, McLean thought, would provide a useful test of the airframe's stability. They designed a simplified quadrant seeker, without a gyroscope, which would turn the canards to aim the missile toward the quadrant in which the sun appeared. It was simply an attempt to discover whether the servo moving the canards would work properly. Several Sunseekers were fired without any real advance in information, according to Wilcox.

The Aero 1 zero-length launcher was a contentious issue. McLean wanted a launcher without a rail; the missile would simply clear the launcher without the rail's normal initial guidance. Jagiello thought this was a death wish and told McLean so. If the missile's orientation was not right, said Jagiello, the missile's rear wings might be clipped on launch; McLean, however, was not persuaded. Not until March 1955 did McLean become convinced they needed a better launcher, and then Jagiello and Wilcox had to design one in short order.

Jagiello worried about the lack of manpower compared to the numbers he observed at work on the Hughes Falcon. "They have thousands of engineers and hundreds of aerodynamicists at Hughes working on the Falcon," he observed; "look at what we have here!" Sidewinder had about thirty scientists and engineers and about a dozen technicians, with one aerodynamicist (and assistants).[15] Wilcox always reassured him:

> To Jagiello's worries, my answer was always the same. I explained that, in the first place, we had the advantage that we, being few, could quickly talk together and solve our problems. . . . In the second place, our single aerodynamicist was more effective than were their hundreds because our man was working on a good basic design whereas their hundreds were not. And in the third place, we had a "secret weapon" named McLean who was able to solve about 95% of all the problems he had ever seen within a few minutes of learning about their existence.[16]

Jagiello was not reassured. In early 1951 Joe Jerger at Raytheon called to say he wanted Jagiello to be the aerodynamicist for the Sparrow III missile. Jagiello was fed up with the Sidewinder project and with McLean's ideas. He left China Lake in April 1951 and became Raytheon's chief aerodynamicist on the Sparrow III project. He soon discovered that the Sparrow III's problems were worse than those of Sidewinder. When he compared the Sparrow's radar with the Sidewinder's electronics, he realized the Sidewinder's strength was its simplicity. Jagiello also arrived at Raytheon at a difficult time; Sanders had gone and Raytheon was virtually starting over.[17]

But while Jagiello had given up on Sidewinder, the team had not given up on him. In early 1952 McLean and Wilcox asked him to come back in time for a review to be conducted by the Guided Missiles Committee of the Research Development Board. The group was chaired by Clark Millikan and included Alan Puckett of the Jet Propulsion Laboratory. Jagiello knew Millikan, and Wilcox wanted Lee to be there. Jagiello came back, first to the Ballistics Division and eventually as a consultant in the Aviation Ordnance Department—a shift that had serious consequences for the Ballistics Division.[18]

As for Hughes, Jagiello was not alone in his concern. For many, beating Hughes was a personal challenge. The Hughes engineers viewed themselves as the real missile experts and looked down on the underfunded and unannointed Sidewinder team. The company was proud of its personnel and its excellent manufacturing plants, all of which led to a certain smugness. In January 1955, shortly after L. A. Hyland had become general manager at Hughes, the secretary and undersecretary of the air force confronted him at a meeting: "You guys at Hughes Aircraft think you're operating a country club. You are an arrogant and expensive luxury and you can't continue to get away with it." But in fact they did. Some Hughes engineers discounted China Lake: "Sidewinder was a little operation no one was paying attention to," according to Simon Ramo, then in charge of the Falcon project at Hughes.[19]

[87]

Tom Amlie observed that Hughes engineers had much higher salaries than their navy counterparts, drove fancy cars, and sported elegant suits. The differences in lifestyle were symbolized for Amlie by a spectacularly attractive woman named Ginger who was in charge of the Hughes antenna laboratory. Ginger drove a Ferrari. "We had nothing like [Ginger] at China Lake," he noted wistfully.[20]

Sooner or later, it always came back to the same questions: Was Bill McLean's design good enough? And could the Sidewinder team's dedication compensate for its modest funding? In Washington the navy knew it must soon decide.

Admiral Parsons Makes a Judgment Call

In May 1951 McLean applied for navy funding to move the project from the exploratory phase into development as a fleet weapon. The Research and Development Board, however, was not impressed. It noted both a lack of documentation of results and "unrealistic fiscal information" about past performance and future plans. That fall, while Sidewinder was still on marginal funds from BuOrd's fuze group, a visit from Adm. "Deak" Parsons finally propelled the missile into development.[21]

Parsons was due at China Lake on 11 October 1951 to make BuOrd's decision on Sidewinder and OMAR. Realizing that this was a crucial turning point, Cdr. Tom Moorer, China Lake's experimental officer, met early in October with navy officers attached to the Guided Missiles Committee of the Research and Development Board, as well as others from BuOrd and the Office of the Chief of Naval Operations. He noted several problems with the officers' perceptions of Sidewinder and OMAR and recommended ways to present the subject effectively.[22]

The fateful day arrived. Wilcox, who had not expected to attend the meeting, got an urgent message from McLean to join him in the packed Michelson Lab conference room. McLean was standing with his back to the blackboard at the foot of the table. L. T. E. Thompson and Parsons were sitting together at the head of the table, facing McLean across its expanse.

Wilcox recognized Parsons, "a tall, lean, and handsome figure of a naval officer." Both men had been at Los Alamos during World War II. Parsons was famous for his role as weaponeer on board the B-29 *Enola Gay* when it dropped the atomic bomb on Hiroshima. He had helped shape China Lake. Now he was deputy director at BuOrd and a representative of the Research and Development Board, which had to approve technical projects.[23]

At this point, Sidewinder was hardly a success. The SCR-584 radar-mounted detectors and the tripod-mounted seekers showed that tracking was possible, but there was no working missile. The Sunseekers had proved little. Still, it was evident that Thompson had confidence in McLean. And so, after listening to McLean, Parsons announced that the Bureau of Ordnance would give the Sidewinder project $3.5 million for the current fiscal year. China Lake would get full technical direction and authority for the project. The transfer of authority was very unusual and effectively took the project out of Washington's hands and gave it to China Lake. Parsons stipulated the requirement for a weekly "speed letter" progress report to the staff in Washington; previously, communications had been few and far between. Wilcox got the speed letter job. The project was on.

The Navy Notices

Washington started paying attention. The fuze funding had been largely invisible because fuzes and their countermeasures were two of BuOrd's most classified subjects. In particular, the project had gone unnoticed by BuOrd's Guided Missiles Section. No longer. The section had a two-person office in Pasadena, so it sent Lt. T. J. "Jack" Christman out to China Lake to find out what was going on.

Christman, one of just thirty-seven engineering duty only (EDO) officers in the navy, had a degree in electrical engineering from MIT. Trained to manage technical development activities, EDOs ordinarily were not expected to rise to flag rank. Cristman decided he would be the exception; and he was. What he did at China Lake does much to explain how it happened.

He made trips to the desert and wrote regular reports on the Sidewinder and OMAR projects, both of which were in a fairly primitive state. In short order, Christman became BuOrd's project officer for Sidewinder, which meant he monitored the development even more closely.

His reports impressed McLean, who invited him to come up to China Lake and join the team. Christman, who by then had realized [89] that important things were happening, accepted. He decided his job would be to provide bureaucratic cover for the Sidewinder project, and he quickly developed project reporting into an art form. He wrote his superiors so many reports it was hard for them to keep up. McLean insisted on seeing his cover letters but never second-guessed him.

China Lake had the conn, but BuOrd had the questions. Christman numbered and catalogued them and made sure each one got answered. He used appendixes to add technical detail in profusion. BuOrd felt that it was being properly informed, yet found it difficult to exercise any real control. Christman's writing bought the project precious time. Having worked for two years with the BuOrd end of the relationship, he knew when to request money and how to assure that funding was timely.[24]

The team tried to draw him into engineering, but he resisted. "I knew the job had to be done, and I knew that if they didn't get their information, Bill wasn't going to get his money, particularly once it came into the clear, and the fleet wasn't going to get the Sidewinder missile." So while McLean and his troops happily designed and tested, Christman protected Sidewinder's financial flanks against the bureaucracy.[25]

He also asked young Lt. (jg) Tom Amlie to look after the enlisted men, and get them housing and bank accounts. Amlie soon became the project's socioemotional leader for the enlisted men, often taking them out for beer-and-pizza parties and acting as a peacemaker when civilian and military technicians' hostility got out of hand. Sometimes he used his considerable physical prowess to knock heads together "until the men calmed down." He also acted as advocate for the men against the more straitlaced members of the team. When Walt LaBerge wanted suggestive pinups taken off lockers, Amlie intervened. "This is the only successful guided missile laboratory in the world; let's not screw around with it." The pinups stayed.[26]

McLean's Design Philosophy

McLean demanded simplicity, low cost, and user friendliness. Many engineers want the same things, but McLean was relentless. His World War II experience with fuzes and arming devices at the Bureau of Standards had taught him the value of simplicity. Low cost followed naturally if the design had not been cluttered up with unnecessary parts.

Developing a simple solution came from following many trails and picking only the best. Flexibility was part of this approach. One of his basic teachings was not to get stuck on knotty technical problems, which often delayed designs unnecessarily while engineers wrestled with brainteasers. "Wire around it," he would say. "If there's a mountain in the way, don't try to go through it, go around it!" Doug Wilcox (no relation to Howie) watching the Sidewinder team, observed that McLean's first inclination when confronted with a problem was simply to get rid of it.[27] Wilcox later put this to good use when he and McLean were developing ASROC, a rocket-launched antisubmarine torpedo:

> McLean [would say]: "Where is the problem? Is it the connectors? Well, do away with them." The first torpedo we had, had a hydraulic system to . . . steer the thing. We had all sorts of problems with hydraulic seals, pumps, and so forth. And Bill McLean told me, "Don't try to be a hero; and just because you run into a technical problem in the engineering, don't let your ego get ahold of you, where you are going to build the best pump, or the best O-ring, or the best thing in the world, to make the thing work—which is normal for engineers to do. But wire around it." Well, it turned out we didn't need to do it hydraulically. So he said, "Well, instead of trying to solve those hydraulic problems, and O-rings, and if you don't need the hydraulics, just design one that doesn't [need hydraulics]."[28]

McLean wanted simplicity because it would lead to mass production and lower costs—which in turn meant the aviators would be able to fire more training rounds. But not everyone was prepared to go as far as McLean. One day four Philco engineers had arrived to assist McLean in producing the first A heads, and one referred to buying AN bolts (AN for standard "army-navy" nomenclature). McLean had no idea what they meant, and after Jagiello pointed out that they were the rigidly specified bolts used in aircraft, McLean said, "No, we want to use the kind you get from the dime store!" The Philco engineers were upset, but McLean prevailed. Falcon and Sparrow were designed without such emphasis and cost much more than Sidewinder to develop and produce. Philco came to appreciate the simplicity of the original Sidewinder design and the thought that had been given to its "producibility." In time, however, even Sidewinder became a very expensive missile. Jack Rabinow joked that McLean had designed the missile so it would be worth its weight in silver. When it was redesigned, it was worth its weight in gold.[29]

[91]

But what most concerned McLean was that the system respond to the needs of the user, which meant the designer had to have direct contact with the ultimate user.

If our designer is to be truly successful, he must have a more direct contact with this consumer than can ever be provided by a set of written specifications. His first task is therefore to get out in the field and get clearly in mind the functions that the consumer would like to perform. . . . The designer who does not take the trouble to try to broaden his specifications by understanding the basic problem will seldom deliver an outstanding product. . . .

In cases where the designer does work from stated specifications they can quite often be misleading. During the Korean war an urgent requirement was received for an antitank warhead capable of penetrating 11 inches of armor. Since we knew it would be impossible to fire perpendicular to the armor under all circumstances, we took a nominal value of 60 degrees for the obliquity of penetration and designed a shaped-charge warhead capable of punching a hole through 18 inches of armor. This weapon was delivered to the operating services in great haste. Some of us became curious as to the motive power employed by Russian tanks that would enable them to run around over rough terrain with armor 11 inches thick. We found that the actual armor of the tanks had a thickness somewhere between three and four inches, and the specification given us had resulted from the correction for obliquity having been made twice while it was coming through channels. *It is essential for the designer to ques-*

tion his specifications and to go back to primary sources in order to develop a real understanding of his problem, and the basis for the need, if he is to create a successful product.[30]

McLean felt sorry for the designers at Hughes, whose Falcon had a rigid specification that had forced them into a technical straitjacket. Sidewinder, on the other hand, reflected McLean's philosophy. The missile was almost a stand-alone system; it could be attached to nearly anything and pointed at the target. The missile would launch whenever the pilot punched the button. Falcon was part of a vast integrated interception system, and much control rested with the automatic system. Falcon would not fire unless the fire control solution was perfect.[31]

Moreover, Sidewinder was a true "fire-and-forget" missile, while most radar missiles were semiactive, which (as has been noted) required the firing plane's transmitter to keep bathing the target with radar to reflect back into the missile's antenna. Sidewinder homed without any further intervention, leaving the shooter free to maneuver. Pilots *liked* fire-and-forget.

Washington did not miss these points. Adm. M. F. Schoeffel, chief of the Bureau of Ordnance, commented in a letter to Capt. P. D. Stroop, then commanding China Lake, that he was aware of the thought that had gone into Sidewinder: "I am personally much impressed with the fact that some really profound thinking has gone into the simplification of the mechanism, thinking of the same high level as was evidenced by Mr. Norden in his design of the Mark 15 Bombsight. Such profound thinking is the very antithesis of the empirical, 'gadgeteering' approach that is plaguing so much of the national research and development effort and making it so impossibly costly."[32]

Sidewinder was coming together. As the team sweated out the various seeker designs and other details, questions were getting answered. The seekers, rocket motor, rollerons, torque-balance control system, and generator were undergoing tests. All the parts worked, more or less, but would they work together?

[8]

The Painted Bird

Getting It Together

With major funding the Sidewinder development gained speed. Sub-systems were tuned up and started to work. Efforts to complete the seekers increased. And all through the team came the thought: flight tests were next.

Twenty-two-year-old "Woody" Woodworth arrived at the program in 1952 as part of his junior professional rotation to find the Sidewinder team "very self-contained." With Wilcox's documentation to assist him, he was soon helping assemble a B head. The B head was tracking airplanes better than the A head at that point. They ran the B head down the Baker 4 track and learned that it could handle rapid acceleration to supersonic speed. For some reason, however, its airborne behavior was erratic and telemetry failed to explain why.[1]

The Aviation Ordnance Division, now a department under McLean, was working on several projects simultaneously on various floors in Michelson Laboratory tower. On the first floor the main team was putting heads and missiles together; Bob Blaise was checking out the assembled missile prior to firing in a loft overhead. On the third floor John Gregory and his team were working on the Mark 8 fire control system, and Henry Swift was working on the Mark 16 fire control system on the fourth floor. While the Sidewinder effort would make air-to-air fire control for rockets obsolete, air-to-ground fire control as yet had no solution so the fire control work continued. Eventually it would lead to the Walleye glide bomb. McLean ranged freely, keeping himself apprised and making suggestions. He maintained a demonstration room for visitors with Sidewinder components laid out on a black cloth.[2]

The cross-fertilization of these programs is hard to trace now, but a great deal of information exchange took place. For instance, it appears that Swift's group discovered that magnetic amplifiers could have infinite gain. This was important for Sidewinder, in which "mag amps" were used extensively.

Testing

Development is a much bigger job than research. It involves many more people and much more abrasive contact with a resistant real world. It is a complicated business, involving many iterative cycles of design-test-redesign. Guided missiles are actors who have to play before increasingly demanding audiences. In the beginning, the task is to learn whether the missile will work at all. As the system gets better, the tests get tougher and are conducted by people less friendly to the system and more intent on finding out the missile's weaknesses. Operational tests are conducted by outside groups trying to get the missile to fail.[3] Sidewinder went through a series of such hoops:

1. Feasibility testing of key components

2. Prototype testing, with more features added each time

3. Development testing of preproduction rounds

4. Bureau of Ordnance testing of production rounds

5. Operational evaluation

Data from these experiments paint a picture. Does the system work at all? Under what conditions does it work? What effects do modifications have? Which subsystems work and which don't? In the end, the picture reinforces or questions the experimenter's mental model of what is going on.

Adequacy versus speed of tests is often a trade-off: Tests that produce complete information take more time. Conversely, rapid information gathering may be too superficial to provide an adequate data base for the next step. With Sidewinder, the data collected were usually just good enough to allow decisions for the next step.

Sidewinder was now a substantial piece of hardware, about nine feet long, the rocket tube about five inches in diameter, and the whole round weighing about 175 pounds. The configurations tested varied from the longer EX-0 airframe with square canards and fins to the shorter and more streamlined EX-1.

Improvisation

In the early 1950s Sidewinder R&D took a backseat to the Korean War. Much of the equipment available for testing was old and had been cast off by the navy or the air force. The planes often had been scrounged from operational units. Sidewinder remained an unofficial program that did not respond to any operational requirement. Unlike the Terrier surface-to-air missile being tested at China Lake, or the Falcon and Sparrow missiles, Sidewinder was an orphan. While the Sidewinder group contained excellent scroungers, the program's shoestring nature could not be ignored. As late as 1954 the chief project pilot at Armitage Field, Lt. Cdr. R. Vancil, complained to Newt Ward: "I cannot reconcile the inadequate shop facilities and talent in use at hangar #2. Your great white tower at the lab, with its impressive facilities and smattering of talent and brains, seems to funnel its ideas (when investigations applicable to aircraft are concerned) to a shoe string outfit that is improperly equipped to make hardware out of your ideas."[4]

The team pressed on. Bob Blaise orchestrated the flight tests and James Heflin was in charge of scheduling project planes and pilots. At first no planes were officially available for the Sidewinder tests, but Heflin noticed that propeller-driven AD-4 Skyraider pilots conducting torpedo tests would launch at dawn for San Clemente Island off the California coast and return to China Lake early in the morning, after which they were through for the day. Heflin managed to appropriate them for the balance of the day. Fitted with Sidewinder launchers, the AD-4s conducted many of the flight tests, until August 1952, when the project got a two-seat F3D-1 Skyknight twin-engine jet.

The Skyknight was "one tired plane," according to Cdr. Glenn Tierney, one of the test pilots who flew it; getting up to forty thousand feet took an hour. Wally Schirra, essentially the first Sidewinder test pilot, also flew the F3D. His logbook records that he checked out in the F3D-1 on 25 August 1952, and the same aircraft was used for most of

Early Sidewinder EX-0 airframe (note square canards forward and rollerons on rear fins). Courtesy of China Lake Naval Weapons Center

F3D Skyknight with EX-0 airframe. The F3D was "one tired plane," but it was crucial for early tests. Courtesy of Robert B. Blaise

the 1952–53 flight tests when Schirra was flying. Even this plane had to be wrestled out of the grip of the Fire Control Division, which was none too happy about releasing it. In December 1954 the Sidewinder project got F9F-8 Cougars; finally, in 1955, it would get its first supersonic jet, an F-100A Super Sabre on loan from the air force.[5]

An esprit de corps filled the project. People who understood the value of teamwork filled key positions. Wilcox and LaBerge were confident and capable. LaBerge often entertained the group with his doggerel verse, usually to the tune of popular songs. "On the Road to Inyokern" (based on the musical adaptation of Kipling's "On the Road to Mandalay," made famous by Frank Sinatra) became a staple. McLean, Wilcox, and LaBerge spent a considerable amount of time ranging the project, inquiring, directing, and encouraging. This personal leadership did much to inspire and motivate, as well as provide important hands-on technical direction.

A Climate for Ideas

Outsiders called China Lake "Bill McLean's Hobby Shop." It was not a friendly jibe. It represented a deep hostility by conventional practitioners toward what they regarded as unsystematic tinkering. Nor were such attitudes held only by outsiders. Albert Christman, in preparation for writing the early history of China Lake, interviewed one of its R&D managers in the early 1970s and found that this manager regarded Sidewinder, Shrike, and Walleye as "accidents." Yet the three systems, all extremely important, arose from careful study of the needs of weapons system users and a highly efficient, if unusual, development process.[6]

For example, McLean took sketches directly to machinists and might get a part in half a day—and test it immediately. This can be contrasted with standard procedures, which might take three days. He was casual about documentation before the fact but insisted on precise documentation afterward. What the machinist had actually made had to be carefully recorded, because it was what had gotten tested. This approach saved time. But more important, the direct contact between the head of the project and technicians resulted in fewer communication hurdles and inspired the technicians, who subsequently went the extra mile to get things right.

A flight scheduling meeting. Cartoon by Howard A. Wilcox

Getting missiles ready for flight test. Cartoon by Howard A. Wilcox

Creative reshuffling of resources was the rule. Blaise recalled playing golf in Bethesda, Maryland, with Walter Kennevan, the BuOrd comptroller for China Lake, who often complained of the station's irregular methods. But Kennevan knew the score. On one occasion, asked if certain funds could be used to create a golf course, he responded, "Of course not!"—but added that he would bring his golf clubs on his next trip to the desert. Shortly thereafter the golf course appeared.[7]

Borrowing funds from wealthy projects to support poor ones was a China Lake tradition during the 1950s, and Sidewinder benefited. During the 1960s similar approaches provided seed money for projects that led to many of the new weapons used in Vietnam, including the television-guided Walleye glide bomb.

Today such actions are not only frowned upon but often are illegal. The increasing rigidity of the defense procurement system, which has gone hand-in-hand with less effective systems and higher costs, has almost eliminated such common-sense bootlegging.[8]

Test Pilots

China Lake was lucky in its test pilots. McClung may be exaggerating when he states that they were mostly of Blue Angels quality, but many pilots did go on to become well-known figures. Wally Schirra and Tom McElmurry became astronauts—and Tom Moorer became chairman of the Joint Chiefs of Staff.

Some had to be convinced. When Schirra arrived at China Lake in 1951 from a combat tour in Korea, he thought the tour would be a waste of time. His father had been an ace in World War I, and Schirra had flown ninety missions in Korea. He wanted to go to Test Pilot School at Patuxent River—and he shared his frustration with Cdr. Tom Moorer, the base experimental officer, who gave pilots their specific assignments. Moorer listened patiently, explained that machine guns and rockets were becoming obsolete—and then gave the young pilot his first glimpse of the next generation of aerial ordnance:

Moorer walked me across the street from his office to a large complex of test laboratories. A couple of Ph.D. types were in a room, and they had a dome-shaped device, made of glass, that would fit on the end of a five-inch HPAG (high-performance air-to-ground) rocket. Inside the dome was a small optical device, a man-made eyeball. I was a cigarette smoker in those days, and I had one in my hand. As I crossed the room, I noticed that the eyeball was tracking me—or my cigarette, to be precise. I realized that it was a heat-seeker. "This is your new toy," said Moorer.[9]

Schirra was amazed by its technology and impressed with its simplicity. The elegant design of the rollerons "blew his mind." He also developed respect for the Ph.D.'s staffing the project; until then, he had viewed the species with some skepticism. He became the first Sidewinder project pilot. Soon, he was hooked.[10]

His conversion was not accidental. Blaise had learned the importance of pilot commitment while working on the Terrier antiaircraft missile. It was a real problem to get pilots to fly support for Terrier tests, and the PB4Y Electronic Countermeasures Plane was seldom ready to fly. The mechanics and pilots didn't know what was going on with Terrier and were not motivated to do their best. Blaise decided things would be different on the Sidewinder project.[11] Pilots, starting with Schirra, were made to feel part of the team. These efforts paid off. On one test, the airstart system on Schirra's plane was inoperable, which meant he would not be able to relight the engines after a flame-out. "Let's just get it airborne," he said. "I'll bring this damned thing down."[12]

Schirra knew exactly what the group was trying to accomplish and went all out. On a typical flight, one radar would track the target drone, an F6F-5K, and a second would track Schirra's aircraft. Their exact positions were inked in by individual pens on a map superimposed on the radar plotting board table, showing the precise paths to be flown by the drone and Schirra. The drone's autopilot was set on "altitude hold" to permit the controller to concentrate on flying the drone around the prescribed race track, turn smoke on and off, and light the flares or thermite pots at the proper times.

Schirra's aircraft was vectored by the range officer, usually Walt Brisch or Cecil Daley. Because the drone was usually much slower than the firing aircraft, the closure rates were planned and plotted for each test to put them in optimum position for the launch. Given the equipment, it was a difficult task, requiring considerable coordination.[13]

The F6F-5K had at most ninety minutes on station, which left little margin for error. Sometimes several runs in a row would be negated, but the experiment was still continued until the team got it right. Excellent camera coverage was usually obtained both from the aircraft and from the downrange cameras. Later, B-17 drones provided a much longer time on station for the tests.

The pilots did more than fire the missile; they also evaluated the system design. Tierney recalls:

> Tom Amlie and I went up in the F3D on the first captive flight with the missile, and he said, "Now, we've got this nickel-and-dime voltmeter . . . [in the middle of the cockpit] and when that voltmeter shows $1^{1}/_{2}$ volts, that's the right signal for the missile to see and fire." And I said, "You mean the pilot in a flying situation has to take his eyes off his target and look at a gauge to see if the missile, find out if the missile sees the target? That's unacceptable." That's when we started differing. "We've got to get something besides the damn gauge. You can't have a pilot, a fighter pilot in combat, looking at funny little gauges to see if he can fire or not." So he came up with a tone, and it's been used ever since.[14]

The result was an 840 Hz tone in the pilot's earphone, the famous "Sidewinder growl."

The development test pilot's job was to make sure the system actually met the need of the pilot who would use it in combat. China Lake subjected its systems to tests by pilots fresh from fleet aircraft carriers. They did not have to be graduates of a formal test pilot school, but they had to have plenty of operational experience. Air Test and Evaluation Squadron VX-5 at China Lake—the Flying Nightmares—carried on the tradition for many years.[15]

The pilots often proved critical for selling the technology once it was developed. Other pilots were interested in hearing what pilots they trusted said about new planes and new weapons. Unlike contractor pilots, navy pilots were rotated back to the fleet, where they would have to live down any bogus recommendations.

Staging a Shoot

Preflight checkout of missile guidance units often was tedious. Space was at a premium in Michelson Laboratory, so the team commandeered a storage space, but much of the preflight check was done on the flight line.

Getting everyone into position was hard. The missile's flight time was only a few seconds, but it might go several miles before it came down. The birds were painted a bright orange and chartreuse pattern to help cameras record the flight behavior. Getting useful observations meant being able to see the missile from the side, which meant stationing cameras and observers downrange, and keeping them alert. As soon as the missile was ready to go, the telemetry van went zooming downrange, breaking speed limits to get into position. Bob Blaise would make the final decision to launch.

Frustrations could run high, but Schirra and his successors were committed. He knew the importance of getting the attack angle and position correct, and worked hard to get the firing plane in exactly the right position. Blaise was often accused of being a cloud worshipper because he would pray that clouds would not decoy the missiles.

Guided Missile Unit 61

While Schirra was the group's Sidewinder test pilot from the early days up to the first successful shot, he was not the only pilot who tested Sidewinder during this time. Lt. Cdr. Al Yesensky seems to have made the first successful firing. By June 1954 Schirra had left and Yesensky had been joined by Lt. Rufo Robinson. The two pilots, along with their team of technicians and handlers, formed Guided Missile Unit (GMU) 61, which had been organized in 1952, and moved to China Lake on 16 July 1953 to assist in developing the Sidewinder. It also tested a variety of the "black-shoe" missiles at China Lake including Terrier, Tartar, Talos, and Standard. Eventually, GMU-61 also participated in the BuOrd evaluation of Sidewinder, a key step in moving it to the fleet. Yesensky was the first unit commander; Tierney replaced him in September 1954.[16]

Tierney assumed command of a unit with seven officers, two hundred enlisted men, and three missiles—Sidewinder, Terrier, and Tartar. He realized that not much would get done on Sidewinder if he had to worry about the black-shoe missiles, so he wrote a letter to the secretary of the navy requesting that a separate unit be established for the black-shoe missiles. Amazingly, in short order GMU-25 was created for Terrier and Tartar, leaving GMU-61 free to work on Sidewinder.[17]

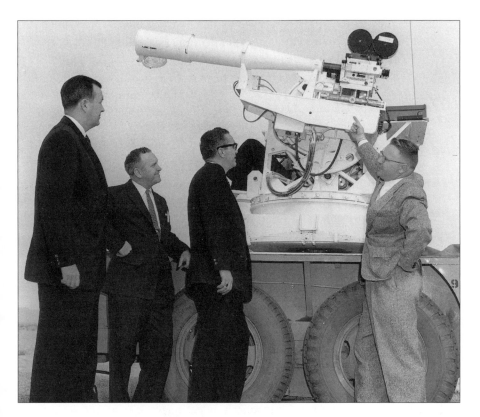

Bob Blaise, *right,* with telephoto camera used to track early Sidewinder flight tests. Courtesy of Lucien Biberman

Tierney's unit, now down to seventy enlisted men and five officers after the creation of GMU-25, had to get up to speed quickly. They were completely integrated with the civilians because their unit shared space in the Mike Lab with them. There was no feeling of "us" versus "them." As the tempo picked up, however, there was too little space in the Mike Lab, and the pilots moved out to the airfield. The sailors (with a new rating, aviation guided missile men) checked out the missiles under the supervision of the Aviation Ordnance Department. Tierney and his fellow pilots ensconced themselves in a nearby Quonset hut. Pilots and missile men were within two hundred yards of the ordnance loading line where missiles were put on the planes.

Tierney had wanted to be a pilot since he was a boy. By the time he was ten, he was selling the *Saturday Evening Post* and reading stories in it about naval aviators at Pensacola. He hung around the local airport, trying to cadge rides from pilots, and when any of the older generation of aircraft carriers (*Langley, Ranger,* etc.) appeared in New York harbor, Tierney would be the first in line to inspect the ship. He entered Columbia University on a scholarship in 1942, but he soon signed up for navy flight training. A series of delays kept him out of combat and "the damned war ended." By then he was commissioned, but he finished up his college work in mechanical engineering. Starting out with an F4U Corsair squadron, he moved through various billets and spent two years as flight deck officer on USS *Kearsarge,* on a tour off Korea.[18]

By 1954 he was flying F9F-6 Cougars on the USS *Boxer* in the South China Sea—and he was bored. An attempt to go to Test Pilot School went nowhere, and then Tierney decided he wanted to go to China Lake. He had been there once and was impressed. Letters through the regular chain of command got him nowhere. Then Cdr. Johnny Refo spotted him in the bar of the Army-Navy Club in Manila. Refo was unimpressed with Tierney's efforts:

> He said, "Jeez, that's not the way to do it. . . . You go back to the ship, see, get out a piece of your squadron stationery with all the stuff on it, you know, and send it to the lieutenant assignment officer, Bureau of Naval Personnel, and tell him you want to go to China Lake. Tell him who you are and what you're doing." In fact, I was kind of the unofficial Top Gun of the Pacific Fleet at the time because of my gunnery averages. So I told him I was the Pacific Fleet Top Gun, and I was out on the *Boxer* in the South China Sea in a Cougar 6 squadron, and I really felt I could do some good at a place like China Lake where they develop ordnance.[19]

One week later Tierney got orders from Blackie Weinel at BuPers to report to China Lake as the officer in charge of Guided Missile Unit 61, to work on Sidewinder. Years later, Weinel—by then an admiral—told him that he had been holding the billet open while looking for an energetic fighter pilot fresh from the fleet. Tierney was in the right place at the right time—in this case, the Army-Navy Club bar in Manila.[20]

Tierney had no prior knowledge regarding Sidewinder but he quickly grasped its significance. Flying F9Fs from the *Boxer* with VF-121, Tierney had patrolled the skies off Taiwan. Ten miles away, Chinese Communists were patrolling in MiG-15s and 17s. The Americans couldn't get their Cougars above forty-two thousand feet, but the MiGs were up at fifty thousand feet. The disadvantage was obvious. Tierney believed Sidewinders could reshape the odds. His job, he recognized, was to get Sidewinder out of the laboratory and into the fleet.[21]

Risky Business

Testing missiles was dangerous. The missile's motor was powerful and could flip the plane over if the missile failed to separate. And its warhead was designed, after all, to tear an aircraft apart. On one occasion, Schirra, flying the F3D, fired a missile that pitched up violently just after it left the rail. The missile performed a series of violent maneuvers, and Schirra thought the missile was looping; he decided very quickly that he had to stay inside that loop. Even though the missile had no warhead, if it hit his plane he would be in deep trouble. Asked years later what it takes to be an astronaut, he said, "You have to think fast!" and mentioned the incident with the looping missile. When asked what he did, he said, "I started chasing it right away!"[22]

But the pilots were natural risk-takers. Tierney once fired two Sidewinders while in a supersonic dive without having been cleared to do so.[23] The California Highway Patrol was not amused by Tierney's love of following Greyhound buses and large trucks along Highway 6 to confirm that he could get heat from the exhaust to register on the missile's seeker.[24] His favorite targets, however, were the diesel locomotives, linked together in threes, pulling big trains through Cajon Pass; they generated tremendous infrared signal. And when the team was joined by air force pilot Capt. Tom McElmurry, the base quickly discovered that McElmurry also had a sense of humor. He liked to hold reveille for the navy by tapping his F-104's afterburner as he arrived overhead at 5 A.M.[25]

Moving Targets

Target drones were an essential test element. Most common early in the program were the QF-6F Hellcat fighters and the QB-17 Flying

Fortresses; also included were turbine-powered Firebee, Regulus, and Matador missiles. Each QF6F-5K Hellcat drone cost about $150,000 in 1950 dollars to convert from an operational aircraft, and it took time to check out controllers—while the controllers practiced, a safety pilot sat in the plane, hands off the controls, during flights, takeoffs, and landings. Even though the Hellcats had their share of mechanical problems, the squadron commander "broke his back to support us," Blaise recalls. Even so, the QF6F-5K drones, based at Point Mugu, never seemed to arrive at China Lake in time to turn around and get airborne before 9:30 A.M. This caused difficulties for the pilots, especially in summer; the air above the base became increasingly turbulent as the morning wore on.[26]

Frustrated, Blaise discovered a number of unused air force B-17 bombers at Point Mugu. The air force team rapidly created a new standard of performance. "The mission briefing the night before the tests was like planning a bombing raid over Berlin," Blaise remembered. "Wheels up" for the QB-17 drone took place promptly at the scheduled 9 A.M., and the bomber could remain on station for four hours, allowing the testing over the ground ranges to evolve at a more leisurely pace. The QB-17s did not need the kind of infrared aids required for the smaller F-6F drones, since the heat from their engine turbosuperchargers was an easy signal for Sidewinder to home on. Eventually, films of one of these B-17s taking a Sidewinder hit on an inboard engine and spiralling down played a major role in convincing Washington bureaucrats that Sidewinder worked.

On several occasions drones went out of control and crashed in embarrassing locations. The Regulus (KDU) missile, as large as a Shooting Star jet fighter, was too complicated and inaccurate to be an effective weapon, but it made a wonderful drone. The missiles were based nearby in Mojave, California, and flew over to China Lake. The drone controllers knew that if the engine stopped, the missile would roll left and go straight down. Jim DeSanto, head of the Range Division, set up a worst-case scenario to establish safe flight paths—and then added a half-mile buffer. And indeed, the worst case did happen when one Regulus lost its guidance signal and crashed safely—half a mile short of a school full of children.[27]

On this particular missile flight, Tierney had been following the Regulus closely at about forty-two thousand feet. When the missile controls locked, the missile went into a dive, and Tierney stayed glued to it. When he realized the missile was headed into the ground, he pulled the plane up abruptly, while the missile exploded into a fireball two hundred yards from the G range tower.[28]

When the Bureau of Ordnance shortly after discontinued the use of QB-17 bombers for target drones, China Lake got its own full-time F6F-5K drone squadron. The new planes dramatically speeded up Sidewinder flight testing.

Crunch Time

What we need is a hit.

L. T. E. Thompson, China Lake Technical Director

Impressions at Midproject

Mickie Benton arrived at China Lake in 1953 with a B.S. in physics fresh from the University of Texas. He immediately entered the China Lake Junior Professional Program, starting with the Sidewinder team. Everything seemed in ferment. Virtually the entire group was located in what is now Room 109 of the Michelson Laboratory. Several types of seekers remained under consideration, and "on every bench there was an engineer with a different design or a different approach." "No two missiles were ever fired the same way."[1] Prototype missile units from Avion, and then from Philco, were constantly reworked in Michelson Laboratory.

Wilcox gave Benton the primer he had written on Sidewinder, and Benton began to sort it all out. While Benton soon became impressed with Sidewinder's *design concept,* it was painfully obvious that the small team was working on a shoestring and that many bugs remained in the system.

Keeping Track

From beginning to end, the Sidewinder testing team used equipment designed for something else and adapted or custom-built its gear from available parts. For instance, not until August 1952 did the team get a chance to use wind tunnels located at the Naval Ordnance Laboratory in White Oak, Maryland.[2] In the meantime, it just had to make do. In line with McLean's design philosophy, however, simple devices were used to provide the data necessary for rapid decisions. "State of the art" often meant "appropriate technology" to record how Sidewinder performed. The major source of data was flight tests. They gave information collected in three ways:

Photographs. Images were provided by the Mark 45 Gooney Bird cameras, developed by George Silberberg, which were mounted on modified World War II gun carriages; the film put observers "on the 50 yard line, 20 rows up."[3] The camera gave the team spectacular pictures, yielding angle-of-attack, trajectories, and accurate miss distances. The missiles were painted garish colors so they were easier to see, "a screwball pattern of orange and greenish yellow," to get roll rate and lift angle.[4] The team also used high-powered telescopes and Askania cinetheodolite arrays.

Debris. Postmortems showed how the rocket motor functioned, or whether the umbilical cord linking missile and airplane launcher had worked properly. Searchers located one servo and seeker unit in a sandy riverbed where it had landed after hitting a drone and breaking in half. They vacuumed out the sand, replaced the broken nose lens, and the unit worked.[5]

Telemetry. Telemetry is the process by which the missile broadcasts what is going on within it via radio signals. While cameras provided information on what was happening outside the missile, telemetry told the team what was going on inside.

McLean recruited Rod McClung to oversee the development of the telemetry system. This included building a van from which virtually all the downrange telemetry was done. While the sophisticated and expensive transmitter units aboard the missile were made by the Seeburg company—famous for its jukeboxes—telemetry receivers in the van were made from Heathkits, which could be found at any electronics shop.[6]

These were assembled by women employees and local housewives, often the wives of scientists and technicians, who learned to read the color codes on resistors and capacitors—crucial when they moved on to building Sidewinder heads—by building the Heathkits. Frank Horton, an electronics technician, supervised them. The women "wired whatever McLean brought in," but especially Sidewinder heads. Eventually, the wiring group started doing work for other projects. The advantage of the Heathkits, McClung noted, was that if anything went wrong with them, they could be repaired easily. The constant tinkering with the gear to keep it working made the engineers intimately familiar with its innards.[7]

Hummergrams

Making sense of the telemetry data was a challenge. There were no off-the-shelf digital recording systems as there are today, and finding ways to display data from the missile's thirty subsystems was hard. A fast camera recorded the data from the screen during the test. But deciphering the film often took days, as the team worked out ways to convert the data from pulses on film to numbers on paper. Everett and Reggie Hill, for instance, were proud of themselves for a system that pulled the film over photodiodes and used signals from the photo tubes to actuate solenoids that in turn depressed the keys on an IBM typewriter, providing an automatic written record. Good but not good enough. Flight tests had already started, and the accurate-but-slow printouts meant delays while the team waited for the results.

Enter the Hummergram, which presented the output as a raster array with all thirty channels displayed, top to bottom, over the flight time of the missile. Named for Bob Hummer, the physicist who developed it, the Hummergram decoded key telemetry information and plotted it against time for instant film reading. The concept emerged somewhere in the middle of the evolution of data display for the project. McClung, whose wise vision shaped the evolution of the system, oversaw the program.[8]

The Hummergram provided a rough but ready record of the behavior of all the missile's systems. "The [telemetry] truck would come flying back from the shoot; the film would be developed and the photograph would still be wet" when the team got to see it.[9] Suddenly the behavior of every part of the missile could be scanned immediately after the test. McClung noted:

The primary value of the Hummergram was just that. . . . Within [two hours] of the shoot, you could hand that record to Bill McLean or his people. They could see pretty much what happened to every channel. . . . And as Bill told me one time, we're actually making a lot of our decisions based upon that. Then he said the actual fine data that comes later, that's accurate down to a percent or two, that went into the report. But he said a lot of the decision-making is being done on [the Hummergram] because you have it so quickly. If the thing failed, you could trace down and see what didn't work. That you could have it within hours of the individual shoot, that was considered unbelievable in those days.[10]

[111]

Data analysis meetings that followed the shots often took place in the late afternoon or evening and frequently required some bold hypotheses regarding what the all-too-sketchy data might mean. Frank Cartwright captured their spirit:

I remember the first time [during 1953] that McLean came down after one of the unsuccessful shots there was a data meeting or something, and I had got my mitts on the [Hummergram], decided where the shot was, the period where this film flight was, and chopped it out, and put that on the table for everybody to look at, including McLean. And everybody was looking at it, couldn't understand it, and Bill got up from the conference table and went over to the trash basket where I had dumped the rest of the film. And Bill didn't say anything, he just brought the piece of film that had the events on it. McLean said "I think this is where the flight took place," and you could see nice signal. I felt considerably chagrined.[11]

McLean's ability to interpret the sketchy data from the telemetry was remarkable; he had an intuitive feel for what the system had done and what needed fixing to correct the problem. The Hummergrams provided the "appropriate technology" that enabled the Sidewinder team to do things other teams found difficult. Like the SCR-584 radar pedestal, they were among the key elements responsible for the Sidewinder's rapid development.

We went fairly directly to free flight tests, in those days, much more so than is done today in many programs. McLean [thought] that free flight test was the real cutter; if you didn't make that one, you weren't going to make it. But if you did make that one, the argument was that the system was fairly simple, so that if you were able to perform in flight, you had passed a lot of milestones that you might be plodding along at in the laboratory. And you might be able to make a demonstration in flight fairly inexpensively that put you way ahead. In fact, that's exactly what happened.[12]

The telemetry van, packed with China Lake's custom-built digital systems, was an innovation. When the team took it to White Sands Proving Ground in 1955, local test engineers wanted them to leave the van behind; no one at White Sands had then figured out a way to digitize the missile output. McClung believes this was the first time that telemetry information was converted from analog to digital form.[13]

The Missile Czar Says "No"

Visitors from Washington arrived almost weekly. These included officials from the Bureau of Ordnance, the Office of the Chief of Naval Operations, the Guided Missiles Committee of the Research and Development Board, and especially the "Missile Czar," K. T. Keller. As all guided missile project managers were aware, Keller's opinion, while advisory in principle, usually became the policy of George C. Marshall, the secretary of defense. A negative opinion from Keller might well kill a project. He was a special hazard for Sidewinder since there was no "operational requirement" for the missile in the first place.

Keller made several visits. On one visit Fred Davis was on the roof using the B head to track aircraft when Keller arrived, obviously in a foul mood, and observed the demonstration. His only reaction was a snort. Sidewinder was still in the feasibility study stage, a long way from production. A decision did not yet have to be made, but Keller decided to make one anyway.[14]

[112]

In December 1952 Keller met with Ralph Sawyer, dean of the Rackham School of Graduate Studies at the University of Michigan, and gave free vent to his feelings that China Lake was "overstaffed, badly managed, and not set up to turn out engineering designs." He expressed similar sentiments concerning Caltech's Corporal rocket effort, but Sidewinder came in for special criticism. "He said it did not sufficiently discriminate against clouds, or have adequate engineering, and that it was not enough better than an ordinary rocket to make it worth while." Keller returned to Washington and sent word to China Lake to terminate work on the Sidewinder.[15]

Fortunately, Guy Suits intervened. Suits, a highly respected figure in the U.S. R&D community, saw the potential in Sidewinder and within forty-eight hours got Keller to reverse the decision. A group headed   [113] by Don Hornig, working on infrared for the navy, had visited China Lake, been impressed, and also pitched in. Against Keller's instincts, the Sidewinder program kept going.[16]

Yet another critical group was the Guided Missiles Committee of the Research and Development Board; it included many eminent civilians and military officers, and visited frequently. Though less powerful than Keller, its individual members were influential. Blaise remembered that "we lived in constant fear that BuAer would take Sidewinder to Point Mugu and kill it. Every missile was going to be the last missile that we fired," unless the committee was suitably impressed. Frederick Darwin, at that time executive secretary of the Guided Missiles Committee, recalls his surprise when the committee first found that China Lake had a promising air-to-air missile: "I remember the visit when we discovered the Sidewinder project. It was completely unknown to DoD. It got authorized because it *had* to be! It worked! That was rather unusual back then. There were gobs of missiles [but they didn't work well]. I recall committee members voicing the opinion that I'd rather have one of those [Sidewinder] missiles than two of anything else."[17]

Coping with Opposition

Most tests did little, however, to end the continuing skepticism in Washington that Sidewinder would work.[18] Management expert Peter Drucker, then serving on yet another advisory committee for the Defense Department, said that both he and the secretary of defense, "Engine Charlie" Wilson, were skeptical of any infrared missile. Drucker's technology mentor, John von Neumann, maintained that all proposals should be evaluated on the basis of "How many impossible things do you have to do?" Sidewinder seemed to be facing too many of these. Could an infrared seeker ever separate a bright target from sky clutter? Most people thought it could not. All in all, Sidewinder did not have good prospects. Drucker recommended cancellation.

Wilcox made several trips to Washington to give his famous blackboard pitches, but these merely annoyed the skeptics. Blaise was told "not to bring that fast-talking professor back here to give any more of those talks." While such opposition made life difficult, it evoked defiance from some members of the team. In a 1957 speech on the occasion of the Office of Naval Research's Decennial Year proceedings, Wilcox stated that "it was also no secret that Mr. K. T. Keller was not impressed with the Naval Ordnance Test Station nor with the Sidewinder, when he 'Kellerized' the Terrier, Sparrow, Falcon, etc. but *not* the Sidewinder. This opposition raised our morale, sharpened our thinking, and kept our costs down—it's too bad every project cannot have this type of opposition."[19]

Others worried about the opposition and with good reason. Even when the project was about to succeed, bureaucrats were ready to cancel it.

The First Hit

One of the missile's most disturbing problems in mid-1953 was a mysterious guidance system shutdown. At some point in the flight path, with the missile guiding successfully toward its target, it would suddenly lose guidance. Washington intended to send some observers out to evaluate. In early June, Jack Christman had a planned vacation abruptly canceled to allow him to write a report that could be given to visitors. In three days he and Wilcox rapidly wrote up all the test firings: what had happened and what they hoped they could do with the missile in the future. The results so far were promising, but whether they would be enough to save the project was not clear. The missile was a technical success, but it needed to *hit* something.[20]

By the end of August the team was at an impasse. A dozen Sidewinders had been fired with no indication that guidance operated after the missile was launched. Wilcox told Fred Davis that something had to be done fast to get the missile to work or "we're going to be cut off." As a result, two missiles identical in design were to be prepared, one by Davis and the other by Marine Corps Master Sgt. Fred Med-

long. The Davis round was fired on 4 September and guidance operated for 2.2 seconds, then failed. The time corresponded exactly with
the time the missile was being accelerated by the rocket motor. At that
point, the motor shut down and the missile began to decelerate from
its Mach 2 speed.[21]

One possibility was that the gyro was "recaging itself." Until the
missile was actually fired, the gyroscope angle was fixed (caged)
pointing straight forward and held in position until after launch. The
original cager was a weight sliding on the post holding the gyro,
spring-loaded forward; the gyro could spin while it was caged but
could look only straight ahead, which ensured that the plane, the missile, and the gyro were all pointing in the same direction at launch.
The cager also protected the gyro during shipping and handling.
When the missile was fired, however, the acceleration forced the cager
back, and two spring-loaded pins latched it into place. Someone the-
orized that deceleration forces as the rocket motor shut down were
snapping the pins, allowing the cager to recage the gyro.[22]

[115]

Medlong's Sidewinder then was modified with stronger caging
pins. On 11 September, one week after the unsuccessful Davis round
was fired, Lt. Al Yesensky took the Medlong round aloft in the F3D
Skyknight. This round did indeed guide successfully toward the F6F
drone and passed close to the tail of the drone. But did it hit?[23]

Only one of the cameras had been in operation, so Christman dispatched his assistant, Lt. (jg) Ken Powers, to Tower 9 for the film. The
Sidewinder group assembled in the conference room, and Powers put
the film into the projector. The film clearly showed the missile approach-
ing the drone. And then suddenly, there it was: the frame showing the
closest approach appeared. Powers stopped the projector on that frame.
As the group scrutinized the picture, however, a small dot of light
appeared in the center of the photograph and rapidly grew toward the
edges. The frame was overheating and burning up; Powers turned off
the projector, whereupon Christman ordered him: "On your way out,
please deposit your balls on my desk!"[24]

All was not lost, however. After some preliminary analysis, Wilcox
decided that, technically, the shot was a hit. He estimated that it had
passed about a foot from the drone's tail. The group was ecstatic—
none more than Powers.

Word spread quickly. McLean announced a wine party at his house that evening. The "small party" quickly grew. At about two in the morning LaV thought that Rear Adm. P. D. Stroop, former China Lake commander, ought to be in on the success. So Stroop, then in Washington, received a call from what was obviously a very noisy party at China Lake. In the hullabaloo, he was unable to make out what had happened. "Get Polly Nicol," he said. "She's got a lot of sense; she'll tell me what's going on." Came the reply, "I can't, she's up on the roof!" For the moment, the pressure was off. Parties continued for several weeks.

Deak Parsons, then deputy chief of the Bureau of Ordnance, was one person in Washington who took special pride in the success. Parsons had arranged the first major funding for Sidewinder. The successful shot was his vindication:

> I am delighted to hear the news on last Friday's SIDEWINDER flight. Hearing the initial account of a fairly wide miss I kept my fingers crossed, but now, with the 1' 11" story [the miss distance], I am convinced that it was full confirmation of the possibilities of the design.
>
> During our many months of waiting for confirmation I have naturally raised the question as to whether our approach and emphasis had been right. I always came back to the conclusion that hindsight confirmed the 1952 decision to "give it a whirl," and all consequences of that decision.
>
> Now I join what I am sure is your fervent hope that in our year of "purgatory" you have met and corrected practically all of the technical obstacles to an in-flight solution.

Congratulations and more success to you,
W. S. Parsons,
Rear Admiral, U.S.N.[25]

More Fixes

The first hit, however, began to seem less impressive after the next six shots failed to score. In October Schirra managed to get a 150-foot miss on the sixth and a 1,000-foot miss on the twenty-third. By New Year's Day 1954, there had not been another hit. The reason for the failures was not apparent. The Office of the Secretary of Defense decided to send Dr. Robert R. McMath out to see how serious the problem was; McMath, a notable astronomer, had been K. T. Keller's associate. If the team could not fix the missile's problems, it was possible the program might be transferred.[26]

Then they fixed it. Chuck Smith walked into an engineering lab where the seeker heads were being tested for their resistance to mechanical noise. Admiral Christman explains:

> Reduction of noise was vital, since with the infrared sensor you're picking up a very small signal, and you don't want mechanical noise interfering with it. Chuck noticed that Fran McCaffrey [an engineer] picked up a rolleron, put a jet of air on it, spun it, and put it on top of the [seeker] head. And Chuck said, "What did you do that for?" "Oh," he said, "this is the best mechanical noise generator I've got." The series of six failed flights all used similar rollerons. Chuck says, "What do you mean? Those rollerons are supposed to be dynamically balanced. They shouldn't have generated noise." Well, it turned out we had changed drawings, you know, getting more efficient, trying to get the production drawings and things like that, and somebody had left off the requirement to balance the rollerons dynamically. Now, there are four of these on the missile, and it turned out that since they weren't balanced, they were all generating noise, and they were just killing the signal into the flight-control system, into the infrared system that was measuring the signals. So what Bill McLean said was "OK, let's take all the rollerons off," which they did.[27]

Careful study showed that the rollerons were in fact causing a vibration of increasing frequency as the motor accelerated the missile—which then began "chasing the random noise signal generated by its tail." Tests confirmed that, at *low* altitudes, the missile performed well without rollerons. At high altitudes, however, rollerons were needed—as the team would soon discover.

Balancing the rollerons helped but did not solve the whole problem. The preamplifier carried a significant electrical field, and any relative motion in it set up microphonic noise that swamped the signal from the detector. Smith finally solved the problem by embedding the preamp in a hard epoxy, so there could be no relative motion, and then painted an electrostatic shield over it to protect against any external relative motion. When this fix was later applied to the Sidewinder 1A, the rollerons could be used.

B-17 Shoot-Down

A successful shot followed on 9 January 1954, this time against a QB-17 bomber drone; Yesensky fired the missile. The Sidewinder hit the QB-17 and bent the propeller. The missile ripped a big hole in the wing, which remained on the plane, held up only by the main wing spar.[28] Warren Legler, in charge of visual observations, remembers the famous 9 January 1954 shot:

> The morning of the shot was gorgeous. I tracked the bird the whole way and as it passed the B-17 I saw a shower of bright specks sparkling in the sunlight. It had homed on the #1 engine and tried to go through the propeller. I suppose it got chopped into little bits, and these were the pieces I saw. . . . I remember sometime later, possibly after I came into the group, a picture of that B-17 in a report of some sort, the #1 prop all bent out of shape. The drone guys managed to save it and got it all down with no particular difficulty.[29]

When the drone landed safely, Wilcox put his head up into the gaping hole and wondered what was holding the wing up. McLean, worried, looked at the damage and asked Wilcox, "Just what does it take to shoot down one of these things?" Wilcox reassured him that they no longer built bombers as tough as B-17s.

The air force brought over another B-17 drone to act as a target, and Sidewinder made an even more spectacular hit on 17 February. Nichols described it:

> Well, I surely remember that B-17 episode. They brought it over here, and they scoffed at the notion that any pipsqueak missile like we had could do anything to their drone. They'd had this drone for several years, and they let everybody shoot at it. It was invulnerable, you know, a B-17. They brought it over here, and said "Help yourself!" And sure enough, [we] clobbered the thing. Down she went, it was a very dramatic thing, with all kinds of cartwheels, flames, and engines flying off. . . . It went on and on and on, you thought the thing would never hit the ground.[30]

Just before the B-17 shoot-down, a Washington telegram arrived that said in effect: "Stop all work on Sidewinder: It has been proven that an infrared homing system will never work." This telegram was tacked up on the Sidewinder bulletin board in Michelson Laboratory. After the films of the B-17 shoot-down were taken to Washington, however, the telegram mysteriously disappeared.[31]

When Dr. McMath arrived shortly after to take stock of the project, the shoot-down film particularly impressed him, and he recommended that the Sidewinder project continue. The films of this event also virtually clinched the argument for Sidewinder in Washington. That year Fred Davis gave a classified lecture to the IRIS infrared symposium in Washington. The response was so enthusiastic that the audience "practically stood up and cheered." Sidewinder had arrived.[32]

[119]

[10]

To the Fleet

We visualized the poor bemittened sailor attempting to assemble the Sidewinder missile on the deck of a blacked-out and wildly tossing ship in the freezing spray of a wintry night.

Howard Wilcox

McLean Becomes Technical Director

On 4 April 1954 McLean was promoted to technical director of China Lake. McLean had taken the job because he felt that one essential part of the technical director's role was to protect the station's creative processes. Jack Rabinow, who knew him well, was concerned: "Why do you want to be Technical Director when you like technical things and you don't like to manage?" McLean said, "Because I know who will get the job if I don't take it. So, I'd rather take the job." McLean knew from experience that some TDs naturally understood the station's informal culture, while others never grasped it. Being in charge would allow him to protect the culture. And he did. McLean had an enormous impact on China Lake's culture of creativity.[1]

The Sidewinder project now was in the hands of Howie Wilcox. This was a good thing. Brilliant as McLean was, Howie was far better suited to getting the missile into production. From the moment of the September 1953 shot, Wilcox had begun getting the missile ready to go to the fleet. On 13 September, two days later, he issued a list of the tasks remaining to be accomplished, along with a schedule and assignments. He established 1 January 1956 as the target date for fleet introduction—and he missed by only two days; Sidewinder began operational evaluation on 3 January 1956, the first working day of the new year.

Refining the Prototype

The missile needed a complete set of production specifications. Jack [121] Christman himself roughed one out and Wilcox proceeded, "under the gentle lash of Christman's wit," to refine it. This was not a job that any creative engineer willingly undertakes, but Wilcox understood its importance and pushed it through.[2]

Specifications or not, Sidewinder was hardly ready for production. On shot after shot, the gas grain that drove the generator failed and the missile lost guidance. Doug Ordahl struggled for months with this problem, whose cause turned out to be the plastic inhibitor film collapsing on the burning grain. Ordahl finally got one that worked.

Fuzes were another problem. The BuOrd laboratory at Corona was responsible for the contact fuze while Eastman Kodak developed the influence fuze. When Corona did not perform well enough on the contact fuze, China Lake took over and did it in-house.[3]

The missile continued to corkscrew after launch. This puzzling behavior usually did not stop the bird from hitting the target, but it had to be fixed. The team originally thought it might be caused by loose tolerances on the gyroscope. This would cause the gyro to precess left during boost (making the seeker think the target had moved to the right). Opposite forces during deceleration would cause an opposite spiral.

Amlie decided to check out this theory. "How much slop," he asked himself, "would there have to be in the gyro to cause this behavior?" The answer appeared to be that the gyro would have to move .026 inches. But the well-made bearing had less than .0001 inch of slop in it. The answer had to lie elsewhere.[4]

Amlie approached William R. "Duke" Haseltine, a China Lake ballistics expert, who suggested the answer—which became known as the angle of pressure theory. As the missile accelerated, the main ball of the gyro, which was rigidly attached to the seeker, pressed more strongly against the ball bearing aft than the bearing forward. The bearing dug into the gyro ball ever so slightly, even though it was made of superhard 52100 bearing steel. The apparent center of curvature of the ball thus moved forward and produced the corkscrewing.

It was a serious problem and it could not be fixed without radical redesign. Corkscrewing ate up kinetic energy, reducing the missile's ability to guide. Amlie and others thought what was needed was a conventional "Scotch cross" gimbal bearing, and Don Steward came up with an excellent design. The new gyro was statically loaded on the bench and it worked well.[5]

Since McLean liked the ball gyro and did not want to complicate the design with an expensive new gimbal, Steward worked in secret. McLean, though, with his usual physical intuition, had somehow figured out, simply from looking at the films and telemetry, what the problem was. So he told LaBerge that the answer had to be a conventional gimbal bearing. LaBerge was prepared. He pulled the prototype out of his pocket. "Like this?" he asked—one of the rare occasions when the team was ahead of McLean. While the original Sidewinder 1s had the ball gimbal, the Scotch-cross gimbal went into the improved AIM-9B version.[6]

Air Tests Accelerate

As the missile improved, GMU-61 tested the successive fixes. At one point they fired 100 missiles in three months. The group flew "virtually every plane in the navy," as each was brought in and wired for Sidewinder. Tom Rogers estimated that he fired 92 Sidewinders during the three years (1955–58) he was with GMU-61. Tierney fired 75 to 80 during his tenure, and Carl Quitmeyer fired 32.[7]

The unit had grown. Schirra had been virtually the sole Sidewinder test pilot until October 1953. Then Rufo Robinson arrived to be joined in turn by Tierney (June 1954), Quitmeyer (June 1955), and Rogers (June 1955). When Cdr. Sel May finally replaced Tierney in July 1957, GMU-61 had three navy pilots, two marine pilots—Capt. Tom Murphree and Capt. Bob Howard—and Capt. Tom McElmurry from the air force, who flew in from Holloman Air Force Base, New Mexico, for tests.

Every effort was made to make Sidewinder "user-friendly." The Sidewinder growl was just one example. The pilots worked hard to identify features that could cause problems in combat, and the missile was designed for easy assembly, disassembly, and maintenance. Wilcox noted:

> In line with McLean's general philosophy, we visualized the poor bemittened sailor attempting to assemble the Sidewinder missile on the deck of a blacked-out and wildly tossing ship in the freezing spray of a wintry night. Consequently, our policy was that assembly of the missile should involve no loose screws or other small parts as well as no specific tools. Moreover, all fins, wings, lugs, etc., should be designed for the roughest of handling. The nose dome was to be covered by a permanent metal cap until after the missile had been loaded aboard its carrying aircraft.[8]

[123]

The team also investigated the effects of carrier launch and recovery on the missile. A "little ricky-ticky hydraulic catapult and arresting gear platform" was installed next to the runway at China Lake to simulate carrier takeoffs and landings with Sidewinder missiles on the planes.[9]

Sidewinder Goes into Production

With the conclusion of the initial development program, Philco began to gear up for full-scale production. In March 1955 the design was frozen and quantity production began. Nonetheless, considerable tinkering was still going on at China Lake. Early Philco units were unsatisfactory and failed in flight tests.[10] Mickie Benton said it was not until the third year he was at China Lake, 1956, that the team stopped tinkering with the Philco missiles.[11]

Wilcox was astonished to hear that Mclean had agreed with Philco engineers who wanted to use connectors. McLean told him that he had allowed the Philco engineers to put connectors on the missiles since the missiles needed to be tested—and the engineers had convinced him the tests could not be performed without connectors.[12]

Philco, however, was not going to be the only company producing Sidewinder. Second sourcing was considered good for security as well as economic reasons. Until about 1954 Avion as well as Philco was thought to be a possible producer of the missile, and the navy even considered helping tiny Avion set up production facilities. The Sidewinder team was very impressed with the engineering work done by

Avion and felt that the small company had made important contributions to the missile's design. In the end, Avion decided it did not want to enter the manufacturing lists.

The navy still wanted a second source. In 1955 it solicited bids from some twenty-five companies on a pilot production run of 200 guidance-and-control units. General Electric won that contract in April 1956. In mid-1956 both GE and Philco were invited to submit bids for the production of 12,000 units over the entire 1957 fiscal year—the low bidder to get the majority of the order, but the high bidder to get a "misery" contract for the rest. GE came in with the low bid at $2,600 per unit, about $200 lower than the comparable Philco bid, which meant GE ended up on top of a 60/40 production split for what had increased to 13,000 units.

Philco then found less expensive ways to produce the missile, and its fiscal year 1958 bid was $1,750—$550 lower than GE's bid for that year; so Philco's share became 70 percent. By 1959 the price had dropped to $1,400, closing in on McLean's initial idea that Sidewinders should cost $1,000 apiece. By mid-1962—seven years into production at "the flattest point on the learning curve" for air-to-air missiles—Sidewinder's price had dropped to a fraction of that for comparable missiles.[13]

Producing Sidewinder was clearly a high-stakes game, however, and the large sums involved attracted many bidders. American industry produced about eighty thousand AIM-9Bs alone, worth roughly $1 billion in today's funds. Later model Sidewinders were more sophisticated and had unit costs several times that of AIM-9B. As Rabinow said, "Bill McLean wanted it to be worth its weight in silver; but they made it worth its weight in gold."[14]

BuOrd Sidewinder Evaluation Program

Before the missile that was now being called "Sidewinder 1" could go to the fleet, two extensive evaluations of the missile were required: the first by the Bureau of Ordnance, the second by the fleet's Operational Development Forces. Wilcox did not want BuOrd to conduct its evaluation at China Lake, because he did not trust the Test Department to carry it out correctly. McLean, however, saw the value in a BuOrd test and insisted that it be carried out.[15]

Cdr. Kenneth Wallace headed the Sidewinder Evaluation Committee that conducted the BuOrd evaluation at China Lake, which lasted from July to December 1955. GMU-61 flew three swept-wing F9F-8 Cougars and an F-100A during the evaluation. Production missiles ready for fleet checkout were shipped directly to GMU-61 from Philco's Germantown, Pennsylvania, facility.

GMU-61 pilots fired thirty-eight Sidewinders at wingtip-mounted [125] Thermite crucibles or flare-augmented F6F-K drones. The launches took place at a rate of about four per week from the toughest points around the firing envelope. One Sidewinder was successfully fired at a Regulus missile and another at a Ryan Firebee drone. The F-100A fired four missiles against flare-augmented five-inch target rockets at an altitude of fifty-two thousand feet.[16]

GMU-61 also flew about seventy-one captive flights to establish detection-and-tracking capability and fuze functioning, using real targets such as B-47 and B-52 jet bombers. Current fighters were tried under all weather conditions and target backgrounds.[17]

Sidewinder passed all its tests, and the evaluation committee submitted its report to the Bureau of Ordnance in early December 1955. On the basis of this report, Sidewinder was released for fleet evaluation to the Operational Development Forces beginning in January 1956. The report contained an evaluation of Sidewinder's probable success in combat: an 86 percent probability of kill if the missile were properly fired. McLean took exception to this, saying it should be more like 95 percent or 98 percent. Both estimates were optimistic. (Actual combat results are discussed in chap. 16.)

Fleet Evaluation

Air Test and Evaluation Squadron VX-3 of the Operational Development Forces, stationed at Atlantic City, New Jersey, was to carry out the test of Sidewinder on board the aircraft carrier USS *Hancock* (CVA-19). Wilcox had promised in 1953 that Sidewinder would be ready for evaluation by the fleet on 1 January 1956, and it was. Tests were to start on 3 January.[18]

Philco, though, was having trouble providing the missiles for the test. A California supplier had sent a defective batch of gas grains. A second shipment of 144 grains was readied in record time and sent off to Philco—but by railway express rather than air express. Tom Rogers and another airman launched in the F-3D for Hill Air Force Base, in Ogden, Utah, to try and intercept the Union Pacific freight train carrying the grains.

After landing at Hill, Rogers took several airmen for a midnight mission in an old air force truck expecting to intercept the train at the top of the Ogden grade. In snow up to their waists, they got to the train, only to be put off and told that "no way are you going to have these things, son. We have an armed guard here. These things are confidential." In the end the air force personnel signed for the servo grains and the shipment, about twice the size of a case of beer, was packed into the F-3D.

Rogers flew to Naval Air Station Olathe, Kansas, where he blew a tire, borrowed one from an F4U Corsair and, after further adventures, finally landed at Willow Grove near Philadelphia, where they met the Philco truck and delivered the grains. Operational evaluation took place as planned.

Checkout Module

Sidewinder needed test equipment, so a parallel crash program was going on to design a seeker-head test system for use on board aircraft carriers. Engineer Jim Madden had designed a large cylindrical checkout module the size of a washing machine. McLean hadn't wanted anything elaborate; he thought a flashlight would be enough to tell whether the head was working. If it wasn't, the head could simply be thrown overboard. But the navy insisted, so Madden's monster was built, with the aid—but not the enthusiasm—of Mickie Benton: "[Madden's] concept . . . was a whole series of tests. It must have been eighteen tests, more than we needed. I could never talk Jim out of the complexity of the system. . . . It had fifty relays at least. . . . I've forgotten how many microswitches. . . . It really qualified as Rube Goldberg, but we made it work, and we built several of them!"[19]

Benton found himself working fifteen hours a day to finish the project on time. It was received with great reservation by naval officers, who at first were not ready to believe it would fit on the dumb waiters taking equipment to the hold. Peter Nicol, however, had carefully measured them and assured his naval counterparts that the washing machines would fit. The units had a high failure rate and in time were replaced by a simpler and more reliable handheld electronic system designed by Benton and Amlie. Another problem was solved when Bob Blaise suggested that special dollies holding four Sidewinders apiece would allow Sidewinder stowage in the ordnance passageways. Such dollies proved quite efficient and were used for many years.[20]

The fleet evaluation of Sidewinder 1 was largely positive. After the evaluation, the pilots were so impressed with Sidewinder's performance that they began to shift the nuclear weapons out of the way to make room for Sidewinders.[21] About 80 percent of the missiles worked; the failures were caused primarily by the difficulties in getting the gas generator grain to function properly. Two other problems remained: an excessive missile roll rate and the corkscrewing motion through the air. The roll rate, which didn't seem to hurt the missile's accuracy at low altitude, was fixed eventually by adding back the rollerons. The corkscrewing was solved by a new gyro. Both fixes took place during 1957, when the Sidewinder 1A replaced the Sidewinder 1.[22]

[127]

Preparing the Fleet for Sidewinder

If Sidewinder had to get ready for the fleet, the fleet had to get ready for Sidewinder. Much of this work was done under the direction of Peter Nicol, whose recruitment showed McLean's ability to draw talented individuals into the project. A chance meeting with Nicol while McLean was testing some underwater swimming gear led to Nicol's joining the team in January 1955.[23]

Nicol's job was to find out what was needed for regular shipboard installation of Sidewinder, a task for which no blueprint existed. Nicol, however, soon turned this inquiry into a fine art. He developed a comprehensive description of the missile and then drew up an exhaustive list of all the shipboard support equipment that might be required to handle the missile, its components, checkout gear, and assembly process. This involved considerable imagination, but Nicol simply continued in his logical way to sort it all out. In the end, he prepared a large document that did much to ready the ships to receive Sidewinder.[24]

To spark their creativity Nicol and Jim Madden decided to inspect a Sparrow I missile installation on the USS *Hancock.* The Sparrow I was larger and more complicated, but the two men expected to learn enough to help them with Sidewinder. On board the *Hancock* it rapidly became clear that sailors were not happy with Sparrow stowage aboard ship. An ordnance handling officer frankly stated to Nicol that he felt guided missiles would not find a home aboard ship.[25]

The two then went to Norfolk Naval Shipyard in Portsmouth, Virginia, and also spoke with Adm. Charles Helm, head of the Bureau of Ships (BuShips), who seemed cooperative. They returned to China Lake, and in November 1955 Nicol prepared a shipboard installation plan to provide all the information possible to BuShips and the various shipyards, making several trips to BuShips to discuss details. They returned to Norfolk and other shipyards to provide the ship designers with information about the required facilities. As Nicol observed:

> This was probably the first time that anyone from [China Lake] had actually gone aboard ship for the pure purpose of putting a weapons systems, especially a guided missile system, aboard ship. Some people probably did it for Sparrow, but it turned out that in almost every case, when I visited a shipyard to brief and indoctrinate them, they were extremely happy to receive me and to receive the information that I gave them, and several of them sent back letters of appreciation, which indicated to me that not many, if any, people had done this previously.[26]

At Norfolk, where the USS *Randolph* (CV-15) was being fitted out to accept Sidewinders, shipboard safety procedures required that electronic components be stowed separately from explosives. Thus the guidance sections had to be put in a different stowage area from the rocket bodies. The warheads had to be stowed in a third place, the fuzes and wings in still other places. All this meant assembling missiles before loading them on planes. Nicol rapidly came to the conclusion that the place to do this was the cafeteria.

Once again, there were problems. To make shipboard life more habitable for the sailors, tables in aircraft carrier messes were being welded to the floor. Nicol asked Charlie Helm for help. Helm said he would take care of it and ordered the tables to be bolted to the deck, which meant that 296 bolts would have to be removed to clear the area for missile assembly. Nicol pointed this out to Helm but was told that

BuShips had done all it could to accommodate Sidewinder; eventually, however, this too was changed. There was simply no time to take out 296 bolts in a combat situation.

Because Nicol always approached these problems in a diplomatic way, he was able to get action, while a more aggressive person might well have created conflicts. The personnel at BuShips quickly learned to appreciate that Nicol would be thoughtful of their needs. Charlie Helm was always responsive to Nicol's requests. Nicol was a living example for one of China Lake's most valuable characteristics—visiting the customer to ensure that weaponeers and users understood each other. It became a China Lake hallmark.

A Missile a Minute

Wilcox thought that ordnance crews should be able to assemble "a missile a minute." This seemed unrealistic to someone at BuShips, who noted in a letter to Nicol that tests had shown that it took *two* minutes to assemble a missile on a test stand. Nicol asked them if they had considered using two assembly stands. The carriers would normally keep twenty-four missiles assembled at all times on Aero 16B bombskids in a Ready Service area.

Getting and keeping Sidewinders together was the next point for action. Timed results of an assembly test on board ship at NAS Alameda suggested that more rapid and secure means of assembly were needed. And flight tests revealed that the individual sections were coming apart on take-off and landing. The team switched to using clamps while sailors tightened screws using pneumatic screwdrivers; this solved the problem.[27]

The fine-tuning of the process that put Sidewinders and ships together owed much to Nicol's diplomacy and tact; the role that Nicol evolved for himself was key to the success of the complex introduction process. He took great pains to help the ship designers with their task of developing missile storage and assembly areas. He studied their needs, got them the information they needed, and showed up to evaluate and praise their handiwork. No one at China Lake told him to do this. He just did it. In the process, he developed a network of contacts who recognized what he had done. His approach won him friends and allies in shaping one technology to fit the other.[28]

The kind of interorganizational bridging in which Nicol excelled rarely gets enough emphasis during systems development. A missile that does not fit into a ship creates problems for combat readiness and shipboard safety. Sidewinder had been designed to be easy to use, but Nicol was instrumental in ensuring that the navy's warships could handle it.[29]

Training for Sidewinder

Sailors and officers alike had many questions about the new guided missiles. Nicol and Madden answered those questions they could, avoiding those that involved classified information, but the sailors still regarded Sidewinders with a certain awe. The missiles did not require maintenance—which may be one reason the guided missile-man rating disappeared. In fact, the empty spaces in the seeker head were filled with potting compound so the missiles *couldn't* be repaired. McLean believed that repairs created hazards and might cause missiles to fail. Nonworking heads were to be returned to the factory or thrown overboard. Clearly, however, coping with guided missiles was going to require trained sailors. During the Vietnam War, the more complicated Sparrows would cause serious difficulties.

China Lake made many efforts to get training for users of Sidewinder. China Lake produced the first Sidewinder pilot's handbooks for the Sidewinder 1 and the AIM-9B, but later manuals were provided by other navy organizations. Nicol, Cdr. Gordon Duncan, and Master Sgt. Mike Wieczerzak made a Sidewinder training film realistically shot aboard a carrier. Unfortunately, the other navy organizations that took up the training function did not always do a good job. Pilots in combat often overestimated the missile's capability and fired outside the envelope. Such problems were not properly addressed until much later, when the navy developed its "Top Gun" training program at NAS Miramar. Compared with the development program, Sidewinder training seemed less important to some. Combat would show just how critical it was.[30]

[130]

Sidewinder Becomes Operational

On 14 July 1956, VA-46, flying F9F-8 Cougars on the USS *Randolph* in the Atlantic, became the first operational Sidewinder squadron

when they got Sidewinder 1s (AIM-9As). They were followed the next
month by VF-211, which flew FJ-3M Furies onboard the USS *Bon-
homme Richard* (CV-31) in the Pacific. The Fury was used for much
of the early introduction of Sidewinders to the fleet, since it was compatible with early versions of the missile.[31]

China Lake sent representatives to the carriers, usually Frank
Wentink and Bob Sizemore. Each squadron had to be separately
trained.[32] By 1957 Wentink and Sizemore were also introducing the
missile into air force squadrons, beginning at Myrtle Beach, South
Carolina, after which they trained the continental air defense squadrons
using the F-104 Starfighter at Tyndall Air Force Base, Florida. Sizemore eventually took Sidewinders to Taiwan; Wentink went with
Sidewinders to the Canadian navy at Halifax, as well as the Swedish
and Norwegian air force. In 1961 Sizemore spent time introducing the
missile to the Royal Air Force and Royal Navy. Ken Powers assisted
the Europeans in developing their own version of the AIM-9B. Fortunately these introductions generally went well. Frank Wentink recalls
that the AIM-9B was "virtually a foolproof missile."[33]

During 1956 some two hundred missiles were tested at sea on the
two U.S. carriers. The missiles apparently demonstrated a single-shot
kill probability greater than 60 percent.[34] This was much lower than the
performance reported for the BuOrd evaluation but was carried out
under more stringent and realistic conditions. The results so pleased
the chief of naval operations that he ordered all fifteen attack carriers
equipped with Sidewinders. By late 1957 nineteen navy squadrons and
seven marine squadrons were equipped. These squadrons flew F4Ds,
F11Fs, and F8Us. Clearly, Sidewinder was an operational success.

About thirty-five hundred Sidewinder 1s were produced.[35] They
were replaced in 1957 by the far better Sidewinder 1A, known variously as AAM-N-7, GAR-8, and finally as the AIM-9B. The AIM-9B,
with a new, more conventional gimbal, rollerons, and other upgrades,
had twice the guidance time of the Sidewinder 1 and superior high-
altitude performance.[36]

Sparrow Is Put in Play

The Sparrow III, which had been undergoing tests at Point Mugu, followed Sidewinder to the fleet. While complex and expensive, it was

designed for longer range than Sidewinder and could handle head-on shots—which early Sidewinders could not—and it was an all-weather missile. Its hefty sixty-five-pound warhead could do a lot of damage if detonated within a hundred feet. It went to the Pacific Fleet in July 1958, carried by the F3H-2 Demons of VF-64 aboard the USS *Midway*.

The Demons were underpowered. The operational readiness inspection and demonstrations had barely started in Hawaii when the USS *Midway* was called to the Far East in response to the second Formosa crisis. As the ship steamed toward China, the F3H-2 Demons, carrying both Sidewinders and Sparrows, conducted blue-water flight operations with tragic results. Three pilots, including the squadron commander, were killed in the Demons during landings.

The Sparrows eventually got the planes designed for them. In 1960, when the McDonnell F-4 Phantom II went to the fleet, it carried Sparrow IIIs. But it would also carry Sidewinders.[37]

Along with Falcon, the longer-range Sparrow III missile was a competitor to Sidewinder. A Sparrow III is shown here mounted on a UH-2C Seasprite. U.S. Naval Institute

[11]

Selling the Air Force

All you do is aim it and push the pickle and Sidewinder does
the rest.

Glenn Tierney, Test Pilot

Major Scheller's Campaign

Sidewinder was a navy missile, and the navy put it in service as soon
as possible. Getting the air force to accept the missile was another story.
Maj. Don Scheller, China Lake's air force liaison officer, did much to
change his service's attitude. Scheller, who arrived in 1954 fresh from
Korea, knew that China Lake was working on several key weapons
systems, but Sidewinder came as a surprise.[1]

Did the air force need Sidewinder? Many air force planes were configured for Falcons or for air-to-air rockets, but Scheller was concerned
about aircraft that lacked these weapons. He thought Sidewinder looked
good. He was well aware of problems with the Falcon; the radar version showed promise, but the infrared version, which he had observed
in testing, did not work well. Its infrared reticle could not resolve a
point target and often chased backlit cloud edges; Sidewinder's reticle was better. Moreover, he thought both versions of the Falcon were
too complicated and too costly.

Scheller began a campaign to get air force interest. The higher echelons did not respond, but some operational units were curious.
According to Vice Adm. Bill Moran,

During the mid-1950s . . . we began to get some inquiries from various different places in the Air Force other than the air force staff or air force headquarters, such as the tactical fighter community. "If they go on Taiwanese F-86s, maybe they'll go on our airplanes." We did quite a lot of work at the major and lieutenant commander level, and maybe a little bit at the lieutenant colonel level, without the air force senior staff knowing much about it until it was done.[2]

The air force even provided some financing. When the project ran short of money in 1956, Lt. Cdr. Wade Cone, in charge of coordinating missile funds, got some air force money to tide things over. Tierney recalled Cone as a "can do" person. "We were constantly running out of money for equipment. Cone would jump on an airplane to Washington, or wherever, and come back with a string of 'magic numbers,' as he called them, meaning funding allocation and authorization." Cone exploited every possibility. He had taught in the Army-Navy Guided Missile School at Fort Bliss, Texas, and prevailed on a former student—then attached to the office of the Chief of Naval Operations—to scrounge $250,000 of air force money. That was a lot of money in 1956.[3]

While Scheller was searching for a larger opening with the air force, he discovered that a General Wade at Wright-Patterson Air Force Base didn't like the idea of having computers set up intercept courses for rockets and missiles. He was a fighter pilot and he wanted a pilot-controlled missile, one less finicky than the Falcon. Scheller suggested that the Sidewinder was such a missile, that it could shoot down MiGs, and invited the general to visit China Lake. There were few Sidewinders to spare at this point, but LaBerge knew that convincing Wade was important, so he authorized the use of two missiles.

Wade arrived at China Lake, got the cigarette demonstration, and listened to the Sidewinder 840 Hz growl. Then Scheller took him over to Edwards Air Force Base, where Tierney had an F-100 ready. Wade was given a choice of two relatively easy tail shots—a certain kill—or two more difficult shots that would help establish points on the Sidewinder test envelope. He immediately said, "By all means, the latter." He was well aware that setting up the firing regime exactly was important with Falcons; he had been briefed that Sidewinder didn't care about the firing regime, and he wanted to see it work.

Wade was an older man and had to be helped up the ladder into the cockpit. Once there, however, he was very much in charge. The first try was to be a look-down shot against the desert background and required an overhead approach. Tierney set up at Mach 1.2 and twenty-five thousand feet. The target was a red F-6F drone with a flare on the wingtip at fifteen thousand feet. The Sidewinder scored a direct hit on the wingtip flare—a sure kill in combat—but the drone survived since no warhead was installed.

[135]

The second, crossing shot, did not go as well. The flare did not ignite, and Wade's Sidewinder got a false lock-on, probably on a sun reflection. The missile missed. Nevertheless, he was impressed with the missile's simplicity and accuracy, and immediately notified his engineering group at Wright-Pat that he wanted Sidewinder tested on an air force plane. But nothing came of this.

Frustration on the Sidewinder team was high when a gift literally dropped in their lap. A navy A-3 Skywarrior on a transcontinental record-setting flight had problems in flight and the crew bailed out over Edwards. The A-3 flew on, and the air force frantically scrambled four Super Sabres to shoot it down using 20-mm cannon. They failed, and the plane fortunately crashed in the deserted reaches of the Panamint Valley near China Lake. Tierney and others reached the scene and found that the aircraft, having pancaked in, was largely untouched by the fusillade from the F-100s; they could find only one bullet hole. Tierney immediately called for a helicopter and a camera. "Let's photograph this thing," he said. "We'll show the air force how much they need Sidewinder!"[4]

Help from Washington

Scheller was not alone. With McLean's encouragement, Wilcox was running his own campaign to get the air force interested:

I put together a large reel of 16-mm movie film showing the Sidewinder making hit after hit against target drones and rockets. And then I . . . went off to show it to the people at Continental Air Defense (CONAD) Headquarters in Omaha, Nebraska.   And indeed the top CONAD people were somewhat interested. . . . But when I went to talk details with their top planning officer, I think a Col. Ross Thornburg,

Glenn Tierney, Sidewinder test pilot and commander of the Guided Missile Unit 61. Courtesy of Glenn Tierney

I found that he had no interest whatever. He explained to me that the Air Force had its own all-weather Falcon. . . . I found the colonel quite deaf to all my arguments. The Air Force Requirement document specified "all-weather capability," and the Sidewinder did not boast an "all-weather capability," so that was that.[5]

Two months later McLean set up a meeting in Pasadena with Wilcox, Charles Lauritsen, and Trevor Gardner. Lauritsen knew Wilcox from the latter's student days at the University of Chicago and at China Lake. Gardner was special assistant to the secretary of the air force and no stranger to the Sidewinder program. In August 1953 he had headed a committee sent to China Lake by the secretary of defense to investigate Sidewinder.[6]

At that time, however, Sidewinder had been having major trouble; now it was flying high. McLean had told his former mentor Lauritsen about their problems with the air force; Lauritsen had briefed Gardner on the Sidewinder's capabilities. Gardner wanted to know more. The four men sat down to breakfast in a coffee shop on Colorado

Boulevard in Pasadena. Since McLean was better at thinking than talking, Wilcox was quickly drawn into the conversation and soon found himself carrying most of the briefing. Gardner asked lots of pointed questions, but McLean and Wilcox fielded all of them.

After two hours, Gardner suddenly announced: "Okay, I'm convinced. The air force has got to buy the Sidewinder system. I will lay on a study contract and tell my analysts what conclusions to come to. But the generals won't budge on that basis alone, of course, so we will have to have a competitive 'shoot-off' between the Sidewinder and the Falcon. Can you handle such a shoot-off?" Wilcox and McLean indicated they certainly could.[7]

[137]

Holloman Demonstration

Within months, the study was completed with the desired conclusion and the air force formally invited the navy to demonstrate Sidewinder at Eglin Air Force Base, Florida. Air force officials knew that the Falcon's frequency-modulation (FM) scanning system would have trouble with China Lake's persistent bright, fluffy cumulus-cloud background—which posed few problems for the Sidewinder's amplitude-modulated (AM) system.

McLean, however, argued that the test should be run at China Lake. The compromise turned out to be Holloman Air Force Base, New Mexico, near the White Sands Proving Ground. Its cloud formations, though annoying, were not as bad as those at China Lake. Tests were scheduled for 12-16 June 1955.[8]

The first shot was scheduled for a Monday, and the team rolled in the preceding Friday. Tierney and Rufo Robinson flew the rocket motors in on their tired F3D Skyknight and an F9F Cougar. Most of the civilians meanwhile arrived with the GMU-61 enlisted men—watched over by Chief Richard F. Walker. With them came the seeker heads, telemetering units, and fuzes. McClung and technician Joe Pray drove the telemetry truck to Holloman.

Gen. Leighton I. Davis commanded the Holloman Air Force Missile Development Center. The highly respected Davis had invented the gunsight used in virtually every air force day-fighter. Thoughtful and open-minded, he inspired considerable loyalty in his subordinates. Under his direction, Holloman—like China Lake—had a free-wheeling culture for technological testing and was a center that fostered innovation. Davis believed in performance, not regulations, and his openness to new ideas played a significant role in the events that followed.[9]

The China Lakers arrived and were met by Col. Buck Buchanan, Davis's executive officer. Chief Petty Officer John J. McManus, "a little black-haired Irishman with mischief in his soul," told Buchanan the team needed some test equipment. Embarrassed, the colonel replied, "Gee, we don't have any test equipment for the Sidewinder. What do you need?"

"Just a flashlight and a Simpson Meter," said McManus with a straight face. Taken aback, Buchanan took them to see the Falcon test equipment, which consumed about forty feet of wall space. The missile moved on a rail from one station to the next, undergoing tests by a battery of equipment. Apparently, it took nineteen technicians to maintain the test equipment. Clearly, the Hughes Falcon was a complicated missile.

The Sidewinder group also got to look at the new seeker for the infrared version of Falcon but were not impressed. They did not know that the Falcon tests had been going badly. Lt. Col. Tom McElmurry, then the Holloman range officer, remembers the many times a large, slow B-17 drone was brought in and then run around in a racetrack pattern so a plane armed with Falcons could take a shot. Virtually all the shots missed. The joke went: "Please bring that Russian bomber around again so we can have another shot at it."[10]

Sidewinder was simpler, but even so McManus had understated the Sidewinder test requirements. Most of the major checks had been done at China Lake; except for a small console to spin up the guidance unit, the checkout equipment had been left at home. The Sidewinder group was traveling light to gain psychological advantage.

The Hughes group, in return, assigned the China Lake contingent a small corner of one of the hangars, where they set up their console on a workbench next to the impressive and obviously high-tech Falcon checkout facility. They were met by a group of white-coated Hughes engineers who offered to store the missiles in a temperature- and humidity-controlled room. Tierney told them the missiles didn't need any special treatment.

Tierney and Robinson then proceeded to lay out the missiles on a mattress in the bed of a pickup truck. They did remove the seeker heads and suggested that they be stored in a secure place. The rocket motors were replaced on the aircraft. The Hughes engineers were concerned: Wasn't the Sidewinder group worried about an impending sandstorm? Whether out of confidence or bravado, the answer was no.

The Hughes people then asked what kind of test equipment would [139] be needed. Tierney gave them the by-now-standard "just a flashlight" reply but added that a ladder and a few other things would be appreciated. He explained that a man on the ladder would wave the flashlight back and forth to check when the missile was operating properly, so the pilot would hear the tone in his ear. This subtle gamesmanship created the right psychological climate for the show that was about to unfold.[11]

The predicted sandstorm temporarily delayed the Monday shoot. Blaise awakened on Tuesday to find his sheets covered with fine sand; Tierney, meanwhile, in true test pilot fashion, had spent the night before the shoot drinking beer across the border in Juarez, Mexico. At Holloman he told anybody who would listen that "Sidewinder works on a radical new principle: it works."

To underscore the team's nonchalance, McClung and Pray rose early on Tuesday, decided they had time to kill, and took the telemetry truck to see White Sands National Monument. This took longer than expected and they arrived back at Holloman at virtually the last moment before the test.

While Tierney and Robinson climbed into the F3D cockpit, McManus continued to wage psychological warfare in the bleachers, where he and Chuck Smith awaited the first shot. He told the Hughes engineers that since the jet drones were so expensive—they cost $100,000 to $150,000 apiece—the pilot had been told to put a bias in the missile so it would score a near miss. This was pure blarney but it created a ready-made alibi.[12]

The first shot was set up to be difficult. The F3D was to fire downward against the bright white desert sand, a hot background in terms of infrared energy. Tierney had given Robinson the left seat because he felt Robinson deserved it for his previous preparation and test work. Robinson leveled off at twenty thousand feet and then rolled in on the QF-80 drone flying below. Although they could hear the tone, it was hard to detect the IR signal against the bright return from the desert sands. Only a few seconds were available to set up the shot. Tierney asked Robinson if he thought he had a good lock-on. Robinson said he did, so Tierney told him "Shoot!" and Robinson fired. The Sidewinder hit the drone in the tail, which broke off. The QF-80 spiraled downward into the sands.[13]

In the bleachers, the assembled observers burst into spontaneous cheering. Not Chief McManus. Feigning disgust, he threw his hat on the ground, saying the pilot had not put in the bias. The Hughes engineers walked away impressed. Still, skeptics wondered if it might have been a lucky shot. No more drones were available at Holloman, but Blaise, networking by telephone, was able to break one loose from the Sparrow III project at Point Mugu.[14]

Scheduled for the next day, the test was to be a stern shot at thirty thousand feet. Tierney thought this was "shooting fish in a barrel." General Davis delayed the firing until four o'clock so he could be briefed, and Tierney told him they were going to use a "slightly tired missile," which had been used for many tracking tasks. Nonetheless, he said, they would get the drone. Davis scoffed at the pilot's confidence, whereupon Tierney said he "would cover any money in this room, that if we get into the firing box, I'll shoot it down." Davis and Buchanan came up with ninety-five dollars between them and laid it on the table.[15]

Tierney climbed into his F9F and taxied out, only to encounter a twenty-minute delay while a crew pulled out all the radios and the parachutes from the drone; the technicians had learned that Sidewinders did indeed shoot down drones. Mickie Benton found it hard to find room to stand anywhere near the hangar. The word was out, and the air force had turned out to watch.

Tierney set up in a racetrack pattern about three miles behind the QF-80 drone. He closed to two miles and had an excellent target tone. Cleared to fire, he pressed the pickle and the Sidewinder went right up the drone's tailpipe. The drone exploded in an "enormous ball of fire," whereupon Tierney rolled into a dive on the general's headquarters, going supersonic in the process, before pulling up into a victory roll. The team took some heat from General Davis. But he was now convinced. What, he asked, could he do to help Sidewinder?

Never shy, Tierney told him they needed higher-performance planes. So far, he pointed out, China Lake had suffered from its association with the black-shoes at BuOrd. While the brown-shoe navy at BuAer was using the navy's high-performance aircraft to develop its own missile, the Sparrow III, China Lake was stuck with older aircraft. The team's F3D, for instance, took an hour just to get to forty thousand feet, and they really needed to work at fifty thousand feet and above.

[141]

Would they like an F-100 Super Sabre, asked the general? Tierney could hardly believe it. Tierney decided to spread the word of Sidewinder's latest success. He sent a SECRET message to China Lake with information copies to "just about every major navy command in Washington that I could think of." It went something like this:

> OIC-GMU-61 sends: Reporting first two Sidewinder kills against QF-80 targets. First round launched in 20-degree glide at 10,000 feet MSL [above mean sea level] against QF-80 at 5,000 MSL. Maximum IR background, looking down at sunlit white sands. Clean separation, normal flight, missile struck and severed horizontal and vertical stabilizers above tailpipe. Target crashed and burned. Second round launched in level flight, co-speed, co-altitude, at QF-80. Normal IR background. Clean separation, normal flight, missile entered tailpipe.[16]

The message went out on Friday. On Monday morning, Tierney learned that more money was coming and that air force participation was on the way.

The Second Holloman Tests

The official air force, however, was not yet satisfied that the Sidewinder could meet its operational requirements and decided to carry out its own high-altitude tests. While China Lake was to furnish the missiles and targets, the personnel and planes would be entirely under air force control. Accordingly, Cartwright, Tierney, and other team members went back to Holloman in early September 1955.[17]

The air force planned to use an F-86F Sabrejet and an F-100A Super Sabre, and the missiles were to intercept targets at higher altitudes than either of the planes could actually fly. The procedure required the pilots to use excess energy to pitch up toward a higher altitude, fire a target rocket, wait two seconds, and fire a Sidewinder at the rocket. The intercept altitude was to be fifty-three thousand feet for the F-86F and sixty thousand feet for the F-100A. Twelve target rockets were used. The Sidewinders had no rollerons because they had proved unnecessary at low altitudes. Tierney was concerned about having air force pilots carrying out tests that included pitch-up shots.[18]

On the first day, the pilots practiced firing only the target rockets. On the second day, they used missiles without guidance units to test the firing regime. On the third day, they fired guided Sidewinders without warheads. Of the six shots (two from the Sabre and four from the Super Sabre), one was considered a miss; the others, from what the pilots could see, were near-misses, close enough to suggest that missiles with warheads would have destroyed the targets.

Two of the F-100A firings were supersonic. The pilots did note a high missile roll rate, which was caused by a lack of rollerons. The instrument data—photographs and telemetry—were not immediately available but were expected to bear out the pilots' observations. These tests included the first firing over fifty thousand feet and the first official supersonic firing.[19]

Unlike the Sidewinder group, the air force did not rush to analyze the films and telemetry. But the air force was impressed with the tests, issued a formal requirement for Sidewinder, and set up a program office under Col. Paul Cool.

When the films came back, however, analysis by China Lake mathematician Pauline Rolf showed that the miss distances were much greater than those reported by the pilots. Walt LaBerge then convened a China Lake meeting to pinpoint the reason for the wide shots. The data showed that the misses had described a helical flight path, and LaBerge, supporting his arguments using equations, put his finger on the problem: no rollerons. There was more discussion. The decision was to reinstall the rollerons. Major Scheller noted that LaBerge was now writing furiously away on a yellow tablet. More equations? No. After the meeting, he discovered that LaBerge had been writing one of his famous songs.[20]

The rollerons could be reinstalled, but the whole affair was very embarrassing both for China Lake and the air force. Cartwright got the unpleasant job of informing the air force, whose Sidewinder project office had grown quite large. He was chagrined at having to deliver the unhappy news to Cool at Wright-Pat: "It was like having to report that, sorry sir, but the Titanic has sunk." Nonetheless, the air force did not rescind its requirement.[21]

China Lake Gets a Super Sabre

Now higher-altitude tests got underway at China Lake using Sidewinders with dynamically balanced rollerons and precision bearings to eliminate noise. As Jagiello said, "We decided that we couldn't use dime-store bearings in the gyro." Thanks to the good offices of Gen. Leighton Davis and negotiations by Scheller, China Lake got the loan of an F-100A Super Sabre and Capt. Ray Brown along with it.[22]

The F-100A was duly rigged up by the Test Department, and in December 1955 Major Scheller and physicist Warren Legler were involved with Captain Brown in a series of tests with the F-100A in shots at targets up to fifty thousand feet.

The air force had sent only one mechanic with Brown, so the Sidewinder civilians found themselves assisting with many of the manual tasks; Amlie packed the drag 'chute. On one test, Brown fired a successful shot, landed, and headed off to the line shack for a cigarette. Within minutes, however, the civilian team had readied the aircraft for the next test. Brown had barely finished his cigarette before technician Joe Wojecki yelled at him to hurry up. On the next landing, Brown took off running for the line shack. Everyone pitched in on these tests. Legler found himself manhandling one of the birds on to the F-100A's missile launcher and recalls the satisfying click as the missile locked in place.[23]

On this series of seven tests with the tuned-up rollerons, the missiles performed well.[24] On 14 December missile number X-156 came within five feet of a target rocket at more than fifty thousand feet.[25] This was scored as the first successful interception at a speed greater than Mach 1. On 26 December Scheller, Legler, and Brown found themselves at Wright-Pat explaining to a much-relieved Colonel Cool that Sidewinder did work at high altitudes after all.[26] Cool responded by giving China Lake the F-100A on indefinite loan.

But Major Scheller knew that even higher altitude tests were going to be required if Sidewinder was going to meet air force needs. It was obvious to him that Sidewinder needed testing by the kind of aircraft the air force was soon going to be flying—high-speed, high-altitude platforms. Fortunately, they would come soon.

[12]

High-Altitude Testing

Somebody had to go up there and put his marbles on the line.
Glenn Tierney, Test Pilot

Tierney Checks Out in the F-100

By early 1956 Captain Brown had completed the F-100 tests. The team had learned a lot, but the GMU-61 pilots wanted to conduct the next high-altitude tests themselves. So in February, Tierney reported for a week's F-100 ground training at George Air Force Base in Victorville, California. GMU-61 would have preferred to use the navy's new F-8U Crusader, but the navy would not make a prototype available.[1]

Tierney decided he could not afford to spend a whole week learning the F-100's ins and outs, so he drove over to George on a Sunday afternoon and presented himself to his instructor—who explained that he needed one week of ground school and another week of flying. Tierney said he was under considerable pressure and had only two days because he had to be back at China Lake on Wednesday to fire missiles.[2]

"I also told him strange and wonderful stories of the Sidewinder, our shooting down QF-80s at Holloman and all that," Tierney later recounted. They compromised: Tierney would take the ground school examination immediately—Sunday—and fly on Monday and Tuesday. Tierney got the examination, the *Pilot's Operation Manual* ("about the size of the Los Angeles telephone book") and also the "master gouge," which contained the answers to the exam questions. In three hours, he read the important parts of the manual, then dutifully filled in the examination using the master gouge. The instructor pilot then dutifully graded the examination, which Tierney passed with a high score. Then the two pilots repaired to the nearby Apple Valley Inn for happy hour and dinner.

Tierney was an experienced carrier pilot and consummate showman, so as he taxied out on Monday morning, he decided to show the air force the navy way. "With all those air force clowns watching me, I rolled down the runway, sucked up the landing gear at 160 knots, hauled the nose up, and rocketed up to altitude."

His landing was equally spectacular. Air force regulations specified entry into the pattern at 300 knots maximum followed by slightly decreasing airspeeds from the 180-degree position through the 90-degree position prior to rolling out on final approach and flaring for landing. The F-100A had no flaps, and its specified approach speed "over the fence" was 170 knots. Once again, however, Tierney showed them the navy way.

He practiced some carrier approaches at five thousand feet to get a feel for the aircraft and found that the F-100A developed Dutch roll—a weaving-like oscillation—at about 155 knots in the landing configuration. Undaunted, he hit the numbers at 450 knots, bled off airspeed in a high-G break, slammed the gear down, and set up for a carrier approach—constant airspeed, constant glide-slope, nearly full power on.

"The Dutch roll was not considered an especially dangerous evolution; the plane sort of 'wallows' along, just short of stall. The main point is to keep power on, don't let the airspeed go any lower, and add some power just before touchdown to ensure a wings-level contact with the runway," Tierney said later. Accordingly, on the last two landings at George, this time without chase pilots behind him, he made a "navy approach, with a little Dutch roll on final, just to make it interesting for the spectators."

Tierney soon found himself in Palmdale picking up a new F-100A, tail number 576. Rigged for Sidewinder, this plane became one of three F-100s that GMU-61 used in high-altitude tests to prove the Sidewinder's capabilities above fifty thousand feet. The F-100A could climb to fifty thousand feet, fire, and land in less than thirty minutes. It was usually possible to run two tests a day, doubling the high-altitude capabilities of earlier test aircraft.

The high-altitude tests quickly increased in pace.

The Matador Shoot

In December 1955 there was another shoot-off with Falcon (presumably version C) at Holloman. Two shots with Falcons, both misses, were followed by a successful shot from Sidewinder. L. A. Hyland, the Hughes Aircraft CEO, recalled:

> I went to the test range for the shoot-off. At Hughes, we were hopeful, but rather doubtful of success on this occasion. The air force clearly recognized the probabilities: The Hughes missiles were to be fired first and the Sidewinders later. Although the Hughes missiles picked up the targets and launched successfully, the seekers did not work properly. My heart sank. . . . The first Sidewinder was launched and destroyed the target; the second was not required. We went back to the drawing board.[3]

Hyland apparently returned in a fury to Tucson where the Falcon was being produced and insisted on a redoubled quality control effort. It was not enough. He decided to stop production and redesign the missile. Work started immediately, led by Alan Puckett, later CEO of Hughes, and continued through the Christmas holidays. The Falcon assembly line was shut down for four months and the workforce laid off. Only in April 1956 was the new version, the Falcon D, ready for production.[4]

In April 1956 some high-level air force personnel visited China Lake. Still smarting from the earlier shoot-off at Holloman, they informed McLean that their infrared Falcons were performing well and that they had rigged an F-100C Super Sabre to fire Falcons (presumably AIM-4Ds) in only about three weeks and successfully fired the missiles. They wondered if the navy could wire an F-100 for Sidewinder that fast. McLean took the bet. It was risky; Sidewinder was under intense pressure and it seemed as if "every missile had to be successful, or it would be the last missile that we fired."[5]

The air force provided an F-100C that the Aviation Ordnance Department wired for Sidewinder in less than three weeks—a job that

Guided Missile Unit 61 pilots. *Left to right:* navy lieutenants H. Carl Quitmeyer, Thomas K. Rogers, Glenn Tierney, and marine lieutenant Thomas Murphree. Courtesy of H. Carl Quitmeyer

usually took months—and the plane, along with another F-100, made it to Holloman with the missiles on time.[6]

The flight tests were conducted 20–25 May 1956 against Matador missiles. The Matador was a surface-to-surface missile, much like Regulus, capable of about 650 mph—a perfect missile target. For these tests, the Sidewinders had live warheads with proximity fuzes. The Hughes pilots had not had much success firing the Falcon against the Matador. To save face, the air force planned to fire Falcons from an F-102 Delta Dagger jet following the Sidewinder shots. Tierney flew one F-100 and a young air force lieutenant, Ray Anderson, flew the other. Anderson had never before fired a guided missile. During airborne checks before the tests, one of the F-100s had a missile tone problem. When the aircraft decelerated, the pilot would lose missile tone; when the plane accelerated, the tone came back.[7]

Chuck Smith found a crushed Cannon plug with a broken seal. He taped the plug connection closed to seal it and tied it away from a hydraulic line that had apparently chafed it and caused the problem. This fix seemed to solve the problem at the time, but it resurfaced later in an accidental B-52 shoot-down near Albuquerque in 1961 (see chap. 14).

Gen. Daniel E. Hooks, who had replaced the team's old friend General Davis at Holloman, was openly skeptical that the inexperienced lieutenant could fire a live missile successfully. Tierney, however, was quite confident the event would be a "turkey shoot" once the plane got into the firing envelope. Because of the mixed experience with Falcon, navy and air force personnel bet heavily on the test.[8]

In the morning, Tierney's plane was loaded up with four weapons: two telemetry Sidewinders, with no warheads, and two with warheads. The telemetry rounds were to be fired first, with a slight bias; a smoke puff would indicate that the proximity fuze had triggered. Tierney launched and fired the telemetry round, which duly emitted a puff of smoke as it passed within lethal distance of the Matador. Tierney then maneuvered for a live shot, which "made little pieces out of the Matador." Technician Bob Sizemore was elated. "I almost leaped out of the [range] tower when Tierney pranged that thing," he recalled.[9]

In the afternoon, it was Anderson's turn to fire with Tierney on his wing. When Anderson said he had the Matador in sight, Tierney replied, "Let me know when you've got tone." The Sidewinder growl came in clearly and Tierney cleared him to fire. Significantly, the second shot did not go up the target's tailpipe. It missed slightly, but the proximity fuze set off the warhead, which blew the second Matador out of the sky. This was a godsend for the fuze people, because the large number of contact hits had not given them an opportunity to evaluate the proximity fuze on a jet target. They discovered that the time constant—the timing of the warhead explosion—was slightly off and were able to correct it.[10]

Then, it was the Falcon's turn. Assistant Secretary of the Air Force Trevor Gardner had declared the radar Falcon operational on 16 March 1955, but the IR version was not doing so well. In fact, the Falcon pilots did not even get airborne. Tierney noted that the two Hughes test pilots, one of them quite well known, were ensconced in the best suite in the best motel in Alamogordo. They seemed reluctant to get in the air with Falcon.[11]

One reason for their reluctance was the performance of the Falcon launching mechanism. The Falcon and the F-102 Delta Dagger jet fighter were supposed to be an integrated weapons system. The integration, however, left much to be desired. The F-102 carried its six Falcons in a missile bay under the plane. The missile bay doors would open, and a "parallelogram" cage with the missiles on it would extend beneath the plane. The missiles would be fired, the cage would be retracted, and the bay doors would close again. In theory, six Falcons would make the plane a powerful weapons system.[12]

Tierney and Blaise remember one demonstration of the mechanism at work at Holloman. An F-102 was on test stands in a hangar. The procedure was to open the missile bay doors, drop the launcher, release an unarmed Falcon, and the missile would then slide off the launcher onto a mattress. In this particular test, however, it was difficult to get the system into working order. After a one-hour countdown, the missile bay doors popped open, the missile slid off the launcher, and the bay doors popped shut while the launcher was still extended. This and similar events probably left pilots dubious about firing Falcons against difficult targets.[13]

During a debriefing with General Hooks, Blaise discovered that the original purpose of the test, "to determine if the navy could modify an F-100 to fire Sidewinders in three weeks, thus matching the Falcon capability" had been misrepresented. The Falcon system, which had been installed on the aircraft in three weeks, fired only a dummy aerodynamic round—a missile with no guidance unit. Wiring the aircraft for a guided IR Falcon would have required many more weeks.[14]

Hooks was impressed with the Sidewinder performance and wanted to employ Sidewinders, rather than Falcons, as safety missiles to shoot down runaway drones. In fact, little more than two years later, Maj. Chuck Yeager's squadron used its Sidewinders to shoot down Matadors that strayed off the firing range in Tripoli. Yeager thought this wonderful target practice.[15]

The Sidewinder group secretly rejoiced in the absolute humiliation of Hughes and its missile.

High-Altitude Hazards

Flying the F-100 above fifty thousand feet required a partial pressure suit—the predecessor to the full pressure suits worn by the astronauts—to protect the pilots from decompression in the event cabin pressure was lost. Getting the suits did not come without pain. Tierney recalled:

> I remember early on, when we were about to get the first of our three F1-100s, I told Wade Cone that we needed about $1,500 for the two suits, and travel, etc. He literally winced and naturally asked the usual questions, "Do we really need them, can we do without them?" I carefully explained to him that, among other things, "everybody" at Edwards who ventures above 50,000 feet must wear the suit. Should we lose pressurization or (ugh!) eject above 45,000 feet, we would literally expand and come apart due to body pressure. In addition to which, in the event of ejection at higher airspeeds or lower altitude (as in high speed, low altitude fuze tests), it greatly improved the pilot's chances of survival. Further in addition to which, we were not going to do any of the above without the suits.
>
> When it came time for Tom Rogers and [me] to get our pressure suits, we went down to Edwards to see how the "big boys" did it. They kept their suits, helmets, etc., in a temperature and humidity controlled room, guys in white coats all over the place, each pilot had a private little cubicle and locker, like professional athletes—very nice, indeed. Once Jim O'Reilly, a top Hughes test pilot at Edwards and Culver City (Hughes headquarters) drove up from Edwards to "help" us set up our own "pressure suit facility," of which we did not have one. He was appalled and I was embarrassed. I tried to explain to him that the project was operating on one borrowed shoestring and that it was with great effort that we obtained the suits in the first place, plus travel costs to Massachusetts (where the suits were custom-made) and the trips to the 80,000-foot high-altitude chamber at Wright-Patterson Air Force Base, where they tried to do in Tom and me.[16]

While the suits were tricky to use, the GMU-61 pilots got used to them. They might save a pilot's life if the F-100 experienced a compressor stall and could not be restarted. Such stalls often took place after missile or rocket firings, when the engine inlet sucked in the exhaust smoke and debris from the round fired.

Compressor stalls or not, the flight testing went on. To avoid stalls, Tierney often shut down the engine and then restarted it. When he

[151]

compared notes with fellow test pilots Yeager and Ivan Kincheloe, they thought he was crazy—but Tierney felt it was the only way to avoid the compressor stalls. The results on the ground were spectacular: "The base would almost stop. People would stop their cars and watch. Here comes this F-100 in contrails, boiling, you know, Mach whatever, from the south. Everybody knew what it was. And you could literally see the coughing in the contrails. So half the people on the base knew I'd flamed out, same time I did. It got to be kind of spectator sport—for them, anyway."[17]

"One Wild Ride"

Tierney was no spectator. One test on G Range nearly killed him. After a shot and subsequent engine flameout, he realized he was having trouble seeing; in fact he had tunnel vision. He found this very funny, but grasped that he must be suffering oxygen deficit (hypoxia). (Later, it was determined that his helmet had failed to pressurize.) His microphone was not working because the high-impedance microphone needed pressure to operate. He selected additional cockpit oxygen; a wing dropped, and he was in a spin. Tierney could focus, but only "as through a tunnel." Lacking oxygen, his engine dead, Tierney had to think fast.

Able to keep only a single instrument in view, he concentrated on the altimeter, which was unwinding as he descended. The pressure suit then activated. He told himself, I've got time to straighten this out, but realized he would have to get out if he hadn't gotten the aircraft under control by twenty thousand feet. He got the aircraft pointed down and was able to use a ram-air turbine to relight the engine. The entire base was now watching the drama unfolding above it.

As Tierney's vision improved, he set up for a downwind landing on runway 3. Remembering a bar almost in line with the runway, he used it as a reference, dropped the gear passing through five thousand feet, crossed Inyokern Road at 170 knots, landed and popped the drag 'chute—which promptly detached. Narrowly avoiding three fire trucks racing out to intercept him, he finally got the plane stopped.[18]

Test pilots thrive on such experiences. Lt. Carl Quitmeyer described his own feelings during test flights: "I considered it to be rather high-adrenalin flying to go up in an aircraft like the F9F-6, fully loaded with pylons and missiles, and fly the aircraft to the outer limits of its theoretical performance just to see if the aircraft remained function-ally intact or whether it might start to become unglued."[19]

The quest for high-altitude capabilities continued. The navy wanted China Lake to demonstrate Sidewinder effectiveness above sixty thou-sand feet. New tests were to be carried out against Pogo-Hi targets—large, ground-launched rockets that attained altitudes well over a hundred thousand feet and dropped their targets by parachute. High-altitude Falcon and Talos testing at White Sands against the Pogo-Hi targets created pressure to show that Sidewinder could operate at sim-ilar altitudes.

So now Bob Blaise needed a new target. The Pogo-Hi rocket nose cone drifting down on a parachute provided a good target for radar—but Sidewinder needed an infrared source. Flares attached to the nose cone were an obvious solution—but how accurate an infrared signa-ture would the flares give off at altitude?

Blaise turned to Art Breslow, who scrounged a surplus deck hangar from the submarine *Sealion,* mothballed at Philadelphia, as a high-altitude test chamber. With the air pumped out, the new facility pro-vided very useful information about how clear the high-altitude sig-natures of the pyrotechnics would be.[20]

The flares gave off a weak signal in the chamber's rarefied air, but thermite (iron oxide and aluminum) pots were better. A special mix-ture duplicated the IR signature of a B-47 bomber; varying mixtures simulated other aircraft. The pots also proved a way of saving the drone population, as Sidewinders could knock the pots off a wingtip and leave the rest of the drone untouched.

Hitting thermite pots on Pogo-Hi targets proved more difficult. On 16 August 1956, Tierney fired at one of the first Pogo-Hi targets deployed at eighty thousand feet. He ran out of airspeed around fifty-five thou-sand feet, stalled, and spun out. It was obvious it would take at least an F-104 for tests at such high altitudes.[21]

The Starfighter Arrives

While tests with Sidewinder proceeded at China Lake, Falcon test-ing continued at Holloman. Capt. Tom McElmurry, U.S. Air Force, shared the Sidewinder team's dim view of the Falcon's capabilities. Repeated misses against the QB-17 drones during the 1954–55 time period soon made him outspoken on the subject—and got him an assignment as Sidewinder project officer on 5 December 1955. By February 1956, as the high-altitude tests with Captain Brown started at China Lake, McElmurry was given full-time duty as the Sidewinder project officer. While Brown was flying the actual missions in an F-100A at China Lake, McElmurry provided logistic support, since the F-100A was based at Holloman.

McElmurry then took seven months to go through the Air Force Test Pilot School at Edwards Air Force Base. (Later he became the Test Pilot School's operations officer and head of its Space Course). When McElmurry returned to Holloman, Buck Buchanan got him assigned to China Lake as a Sidewinder test pilot. Although the assignment was ostensibly temporary, actually McElmurry became a full-fledged member of GMU-61 for eighteen months. Major Scheller made Captain McElmurry virtually a member of his family. And McElmurry brought some hardware with him—a brand new F-104A Starfighter.

The air force had come through yet again. The needle-nosed Starfighter was just out of development, and China Lake's airframe was number 0016, the sixteenth produced. With the F-104, capable of Mach 2 and zooming to altitudes in excess of seventy thousand feet, realistic high-altitude tests became feasible.[22]

Jagiello suggested mounting Sidewinders on the F-104's wingtips. Kelly Johnson, famous head of the Lockheed Skunk Works, opposed the idea because he thought the missiles would cause excessive drag. This proved a rare occasion when Johnson's intuition failed him. In fact, the Sidewinders improved the F-104's aerodynamics by reducing tip losses.[23]

The F-104 could not stay above sixty thousand feet for long, but it climbed like a rocket—thirty-six thousand feet in 1.6 minutes. After all, it had been designed as an interceptor. It was a tricky aircraft to fly, however; eight test pilots, including Ivan Kincheloe, died during its development.[24]

Flameouts were a serious problem. The Starfighter had no high-altitude restart capability, and its pressurization system depended on the engine compressor. When cockpit pressure dropped, the pressure suit would inflate and squeeze an already tightly constricted pilot. To get a relight, the pilot had to get the plane below forty thousand feet.

Vice Adm. Bill Moran recalled a flameout approach in the F-104 from his days as assistant experimental officer:

> We were doing our first firings at very high altitudes—65,000 feet— but something on the order of a half or a third of the time, [the pilot would] lose the engine [and] the F-104 has poor glide characteristics. You would adjust your approach to go through an initial point that was directly over the downwind end of the runway. In the Banshee it was like 11,000 feet; in the F-86, it was 11,000 to 12,000 feet. With an F-104 the initial point was at 40,000 feet. You came down like a goddamn rock.[25]

[155]

Flameouts were a constant problem for the F-104 with the early engines at higher altitudes, and McElmurry had his share; at thirty-seven thousand feet or so, he could restart the engine. In McElmurry's experience, the engine always restarted, and he never had to make a dead-stick landing during the Sidewinder tests. After the fourth flameout, however, he got a teletype message from Wright-Patterson Air Force Base stating that he should avoid any maneuvers that caused compressor stalls. McElmurry ignored the message and pressed on.

"When McElmurry reported 'Flameout, 0736,' every radio in the area came alive until he was safely on the ground," Blaise said. Today, McElmurry denies that such work took special courage. Shutting down the engine in an F-104 at seventy thousand feet and dropping down to forty thousand feet was routine in the Space Course program that he later managed at Edwards. Nonetheless, McElmurry was variously described as having "nerves of steel" or "ice water in his veins," descriptions accurately matched by his activities. Sizemore, who worked for Blaise in the Test Department, described him as "courageous, tenacious, and daring, the ideal man."[26]

McElmurry's tests demonstrated that Sidewinder would work at high altitudes but also confirmed the problems with missile roll at high altitudes. Once rollerons were reinstalled, though, the problem would be fixed. Jagiello improved the devices further when he noted that at higher altitudes, the missiles would pitch and yaw as well as roll. He invented a canted hinge line for the rolleron tabs, so the rollerons damped pitch and yaw movements as well as roll.[27]

With the F-100s and the F-104 on station, the test pace accelerated; there might be two or three tests in a single day. Pilots no longer needed to prove that Sidewinder worked. The aim now was to find out under exactly what conditions it would guide successfully—its envelope. Modifications to the missile's guidance, rocket motor, warhead, and control surfaces required testing. Already, the prototypes of a new model, the 1A (AIM-9B), were being tested.

"Boola Boola"

Close calls also could have political overtones. The "Boola Boola" incident started at one of the Thursday night data analysis meetings. Someone voiced concern about Sidewinder's ability to go through contrails without losing the target. Tierney remembers that questions were raised: "What would happen if you fired a missile at a target, and it went through maybe a small cloud—or maybe more properly a contrail of whatever it is that you're firing at? Would it lose lock, or would it hold what it had and pick it up on the other side of the contrail? We had to find out."

So a test was set up on the Pacific Missile Test Range off Point Mugu using a Regulus missile drone from the Sparrow III program. Lt. Tom Murphree had just checked out in a recently arrived FJ-4 Fury, and Tierney felt Murphree should be the pilot for the test because the marines were using FJ-4s.

> So we schedule a test. Murph is in the FJ-4, and I'm sitting alongside him in the F9F. We were up against a Mach .80 Regulus One. Co-speed—Regulus doing .80, we're doing .80. The guaranteed absolute maximum range of the Sidewinder, probably a Model-1B, was 15,000 feet slant range, co-speed at 30,000 feet altitude. We were going to get behind this Regulus in the firing box and fire a missile out of range, too far back to hit but when it came off the launcher to let the missile drop through the contrails of the Reg, and we were tucked in, now—no warhead. We wanted to see if the missile would hold what it had and pick up the target on the other side of the contrails. Point Mugu said, "OK, so long as you don't shoot down the goddamn Regulus, because you're not supposed to, and it's not yours. Promise?" "Right."

So we sit behind this Regulus. I call Range Control for a range. They give me 19,000 feet—4,000 feet more than the missile is supposed to go. I have in the F9F a very, very good modified ranging radar called the APG-30A, which Wilcox and his guys cranked up so it went out to about 25,000 feet max range. So I called Mugu Range. "What do you have slant range to target?" And they say 19,000 slant range, and I look at my APG-30A radar, 19,000 feet—4,000 feet out of range—and we're following this thing, sitting right on top of the contrail. I said, "Clear to fire, Murph. You've got plenty . . ." and Murph says, "Here we go!" Boom! Off goes the missile, drops through the contrails, and came out the bottom, looked like it was guiding pretty good, and we figured it "liked" the goddamn Regulus. Went right up the pipe. Right in the pipe. Oh, Jesus! Boy, are we in trouble!

[157]

Sidewinder missiles were all different. . . . [But] this goddamn missile that Murph fired must have had all the variables ironed out to where there was almost like that thing had been reeled in by wire. It didn't move; it didn't wiggle. It just headed right for that Regulus. We swung out on the side of the contrail so we could see the missile below, and it was like someone was reeling that son-of-a-bitch on a wire, 4,000 feet longer than the guaranteed longest range. Right in the pipe.

Well, we had a voice code [at Point Mugu] . . . because they didn't have enough range instrumentation on it. So if it guided good and was close, you would say, "Boola." If it hit the target you would say, "Boola Boola." And Range Control says, "069, what luck?" And I say, "This is 069. Luck Boola . . ." and I hesitated about three full seconds and gave him the second "Boola." And this guy says *What?* Is that one Boola or two?" And I say, "We have a *Boola Boola* here." And this guy, the Range Controller, he knew what that meant. "Roger, I've got your Boola Boola."

So I called the ground crew on our unit tactical frequency in the hangar, and I said, "Grab your guys, pack it in, and get the hell out of there, and get back to China Lake." I went back [to China Lake], picked up the phone and called Dick Ashworth [China Lake's commander], and I said, "Captain, you've got a problem. We hosed that goddamn Regulus!" And he said, "Oh no, you didn't do that? Oh, my God!" I said, "We had 19,000 feet range. It was the most perfect missile that ever came off the test stand."[28]

Later, missiles were fired from pitch-ups against much higher targets. On 29 June 1963 Maj. Bob Walker, U.S. Marine Corps, in an F4H Phantom II, fired an AIM-9D successfully against a target at 83,400 feet.[29]

A New Warhead

The quest for a new warhead started with a spectacular failure in front of the navy brass, proving another of Rabinow's laws: the probability of success varies inversely with the rank of the observers.

On 11 January 1956, just before McElmurry arrived with his F-104, a big live-warhead show at Point Mugu was arranged to demonstrate Sidewinder's effectiveness to Navy Secretary James Wakelin, Adm. Arleigh Burke, chief of naval operations, and others.[30]

Tierney and Rogers, in F9Fs, were to fire Sidewinders at an F6F-K drone carrying flares in a familiar scenario. Rogers went first, and both his missiles homed on fragments ejected from the flares instead of the flares themselves. Ground crews then turned the drone around for Tierney to have a go. For whatever reason, the Sidewinder missed the flares, and the warhead detonated over the cockpit. According to Tierney, the explosion "blew the canopy off, blew holes all through the canopy area, through the wings, which had self-sealing fuel tanks, and the goddamn Hellcat kept flying."

Tierney, frustrated, flew over the Hellcat, and rolled his plane upside down to see the extent of the damage. "The cockpit was all blasted apart. But the son of a bitch didn't go down." It was a bad day for Sidewinder, made worse when a Sparrow III knocked down a drone.[31]

So we land, and we taxi right in front of the grandstand there, and Captain Ashworth is standing at the bottom of the ladder about fifty yards from the grandstand where Arleigh Burke and all these guys are, and he says, "Boy, you've got some tall explaining to do! I don't know how I'm going to explain to Admiral Burke why you didn't blow that F6 out of the air."

So I'm in my dirty old sweaty flight gear, and Admiral Burke says, "What happened on that first shot? The second one blew the airplane right away." I [told him that] the missile missed a couple of feet high, which, you know, it seldom does, and the warhead went off right over the cockpit. I said, "Did you notice when I flew upside down over that Hellcat? I wanted to see the damage. That whole cockpit and canopy was blasted away by the warhead, the wings were riddled with these fragments, and the self-sealing fuel tanks. You know how Grumman builds airplanes. They don't call it Grumman Locomotive works for nothing." He said, "Yeah, I see your point. What you're saying is that if that had been a real pilot, it would have killed it anyway because it would have just killed the pilot, which is just as good."

[159]

It was at this point that Burke noticed the 1936 chauffeur's badge that Tierney persisted in wearing on his flight suit—which gave Tierney the chance he had been waiting for. He launched into his speech:

> My chauffeur philosophy is simply that these fighter pilots are only chauffeurs, whose job it is to deliver their ordnance into the enemy. You can roar around at great speeds and altitudes (as the air force was wont to do), fly a perfect flight, make a perfect carrier pass, and catch a "three-wire," but if you don't have good reliable, effective ordnance, be it Sidewinder or guns, and nail the enemy with it, then you might as well have stayed home. The name of the game is ordnance, not the pilot or the airplane, ship, tank, or whatever.

Admiral Burke agreed, adding that the philosophy applied to ship drivers as well as airplane drivers.

While Tierney had saved face, it was clear to Blaise that Sidewinder needed a more powerful warhead than the fragmentation warhead from the HPAG rocket. He decided to investigate just how effective the warhead really was. The test results were disappointing, and it was immediately obvious that the original fragmentation head would not do. "I learned right then," wrote Blaise, "to always prove your kill mechanism before you design your missile. It's what's up front that counts."[32]

Searching for improvements, Blaise contacted Dr. E. J. Workman at the New Mexico School of Mines in Socorro, New Mexico, who was developing a continuous-rod warhead. The warhead consisted of long steel rods wrapped lengthwise around the explosives and welded together at alternate ends. When the warhead exploded, the rods extended in an ever-increasing circle around the missile, eventually to break into individual fragments. One can visualize a circular saw blade expanding outward, at right angles to the line of flight. The previous warhead had projected its blast energy in an expanding sphere. By contrast, the continuous-rod warhead spent all its energy in this tight belt at a right angle to the axis of the missile. The effect was lethal. China Lake's engineers soon developed an expanding rod warhead for Sidewinder.[33]

In its first flight test, Blaise remembers the warhead slicing the drone fuselage in half, just behind the cockpit. It proved effective out to a radius of about thirty feet. McLean loved it, and it became standard Sidewinder equipment.[34]

The Air Force Hi-Fly Program

Blaise played a key role in testing Sidewinder for the navy, and McElmurry, after his China Lake experience, wanted to play a similar role in easing Sidewinder into the air force. At that time there was no official air force funding or authorization for Sidewinder tests; what funds were available were being used to test Falcon. Nevertheless, when McElmurry returned to Holloman, a Sidewinder evaluation program took place under his direction at the Air Force Missile Development Center right alongside the Falcon program.[35]

China Lake gave McElmurry's operation whatever it needed—seventy-five missiles, a complete Sidewinder checkout laboratory, a small contingent of technicians, and a helpful Philco representative—in hopes that the air force might conduct the high-altitude testing that China Lake could not do without more advanced aircraft.

McElmurry started by using large balloon targets designed for Falcon; balloons carrying appropriate target equipment were simply let loose upwind of the base—when they reached about fifty thousand feet, they became targets. The balloons eventually reached seventy thousand feet, and McElmurry estimates that he fired about thirty missiles himself in these tests. Sidewinder consistently missed these targets. The high-altitude balloons were stationary relative to the air mass; the missile had been designed to hit moving targets, and its proportional navigation system could not handle a stationary target in the thin air at such altitudes.[36]

McElmurry now shifted to balloon-launched targets. The usual navy Sidewinder test consisted of firing a missile from the same airplane that fired the target rocket, which consistently gave the missile a shot from six o'clock—clearly an unfair advantage. The air force wanted to check it out on crossing shots.

McElmurry's team—scrounging like the China Lakers—built delta-shaped targets out of plywood (they later got aluminum wings), and fired them from balloon-lofted launch platforms using surplus 2.75-inch Mighty Mouse rockets. This was the Hi-Fly Project. The first successful platform test came on 17 July 1958, followed by the first successful Sidewinder firing against a Hi-Fly target on 5 September. Many successful shots followed; Sidewinder had no problem with a moving target.[37]

[161]

The Hi-Fly target was a much more realistic high-altitude target, but its limited speed and short level or climbing flight time left some of the launch envelope untested. Fortunately, McElmurry was able to draw on the skills of a German émigré, Dr. Richard Vogt, at the Curtiss-Wright Airplane Company in Santa Barbara, California. McElmurry asked Vogt to develop a target that could meet the requirements: hold constant altitude or climb to sixty thousand feet at typical fighter Mach numbers for a minimum of one minute after launch from a high-performance aircraft. After some discussion, Vogt indicated he could do the design work, and he completed it in thirty days. McElmurry ordered fifteen rounds and they all worked.[38]

By 1957 the air force was finally ready to carry out an operational evaluation of the Sidewinder, the Phase Six tests. These went well, and Sidewinder entered the air force inventory.[39] After the AIM-9B, however, the air force would insist on developing its own Sidewinders. These air force Sidewinders did not work well. Not until the model AIM-9L did the air force get a really good Sidewinder again.

[13]

The Creative Dialogue

Reflections on the Weapons R&D Process

Bill McLean believed there was a right way to carry out weapons research and development. His thinking about this subject had started during World War II when he worked with Jack Rabinow at the Bureau of Standards. They had often discussed the difficulties caused by inefficient organizations. Both thought, for instance, that some projects used too many people. While group performance might improve with some additions, getting too big eventually caused performance to decline. As time went on, McLean's R&D philosophy became more articulated through his letters, speeches, and articles. Elizabeth Babcock, a China Lake historian, believes that Haskell Wilson, McLean's colleague, also had a major role in shaping McLean's philosophy.[1]

With the success of Sidewinder, McLean got many opportunities to expound this philosophy. Taken together, McLean's writings form a coherent strategy for successful government research and development.[2] This strategy has six interlinked elements:

1. Provide an environment where unfettered inquiry can take place.

2. Encourage integrated problem conception, experimentation, and testing.

3. Use input from the ultimate users to shape the research process.

4. Encourage competition between teams and laboratories through overlapping jurisdictions.

5. Judge the effectiveness of systems by their performance not paper studies.

6. Produce only what has been shown to work.

Integrating Information

The great strength of China Lake in the eyes of Bill McLean was its free flow of information about needs, systems, and capabilities.[3] The [163] station united several flows of information: between departments, between military and civilian operations, and between the fleet and the laboratory. Further, the "full spectrum" character of China Lake allowed easy feedback between research, design, and testing. Industry could not operate so freely because of legal constraints imposed by the procurement process.

In developing a weapon, information about the needs of users had to come together with information about technical capabilities. In this way, a dialogue between users and designers would be established that would allow the users to draw on the best technical capabilities.

McLean's solution to uniting the two flows was to allow people he called master designers to operate in environments where they could develop their ideas fully—as at China Lake. The ideas developed by these master designers would also be directly exposed to critique from the users. China Lake's strength, McLean thought, was that it acted as a sensitive inquiring system. It was sensitive in its ability to detect needs and creative, fast, and thorough in its ability to do research. Since pilots just coming off aircraft carriers or even actual combat tours participated in weapons tests, feedback tended to be honest and incisive. Similarly, when the station's research showed that a technological idea was bad, it was willing to let it die.

The Value of Competition

McLean believed in competition for R&D. He recognized that overlapping responsibilities might cause conflicts or even lead to duplication. But better systems would be produced, and more money would be saved, McLean believed, by providing options during the development process. If everything depended on a single technical solution from the beginning, what happened if the assigned problem solver did not succeed? Often, one laboratory came up with a solution to a problem that "belonged" to another laboratory. Designating the problem solver simply created "not invented here" problems if the solution emerged elsewhere. This certainly happened with Sidewinder, which had to fight hard to survive.

In a 1961 letter McLean wrote:

> I believe that the only true measure of effectiveness in R&D must be performed on a competitive basis. This means that we need more than one laboratory in each field of endeavor which is important to government operations; or more realistically, we need two or three groups of laboratories, each having a broad scope of activities extending all the way from research through development, testing and evaluation, and limited production of the type needed to provide guidelines for large scale industrial production. Competition between these laboratories, or groups of laboratories, should be encouraged and the records of their accomplishments evaluated.
>
> Another technique would be to back up every crash program, which the country feels it is necessary to undertake, with a non-crash laboratory development funded at approximately 10 percent of the funding for the crash program. It is my opinion that in a very reasonable number of cases the progress on major programs would be materially aided by taking more time for the thought needed to arrive at an integrated design, rather than to arrive at a crash design by putting together "off-the-shelf" components. By funding the carefully considered integrated program at 10 percent of the rate of the crash program, we ensure that the costs of the completed program will be in about the same proportion. We will only need to be successful one time in ten in order to make such a program pay off. My personal opinion is that the thoughtful, non-crash program will come to fruition before the crash program in well over half the cases. Even in those cases where complete success is not achieved, many techniques and components will be interchangeable between programs.[4]

This was McLean's "10 percent solution" to the crucial R&D problem. Because such programs were small, they were ideal for using the talents of the master designer. For this reason, 10 percent solutions often have worked in practice. For instance, this is typically the principle behind the many successes of skunk works operations, those at Lockheed and elsewhere.[5]

The Dangers of Bureaucracy

McLean believed that bureaucracy posed many dangers to good R&D. Just as bureaucracy harmed the information flow, it also imposed rigidity. The proliferation of rules and regulations interfered with the process of research. In his letters to senior officers, McLean regularly complained about burgeoning and often stupid regulations that were choking off intelligent actions. As technical director of China Lake, for example, McLean wrote a letter to Vice Adm. John T. "Chick" Hayward, then the deputy chief of Naval Operations:

[165]

> We find that, from the level of your office, problems are considered from a breadth of viewpoint and overall responsibility which is not again established until the work reaches the laboratories where it must be accomplished. In between, the job is broken down into a tremendous number of parts for financial management and then must be re-collected and re-coordinated before it is capable of technical accomplishment.
>
> To me, this breakdown and subsequent re-combination represents the basic difficulty in our system. It appears to me to serve only the function of providing jobs for a large number of people in Washington.[6]

Not too many years after this letter was written, McLean's former subordinate Frank Knemeyer convinced the Naval Air Systems Command to use just this sort of consolidation. But all such gains were temporary. Years later, as technical director of the Naval Undersea Center, McLean found himself even more hamstrung by regulations than he had been at China Lake.

More serious than the growth of administrative rules and regulations, however, was the attempt of the bureaucrat to plan the creative process. McLean thought this was certain to damage creativity.

In 1960 he published an article in which he lampooned bureaucratic practices concerning R&D. He argued that bureaucracy had caused creative scientists to be an endangered species; therefore, they should be studied before they disappeared. The most incisive part of the article is a list of techniques for squelching genius. To give the flavor, here are a few examples:

> Concentrate on planning and scheduling, and insist on meeting time scales. (New, interesting ideas may not work out and always need extra time.)
>
> Insist all plans go through at least three levels of review before starting work. (Review weeds out and filters innovation. More levels will do it faster, but three is adequate, particularly if they are protected from exposure to the enthusiasm of the innovator. Insist on only written proposals.)
>
> Optimize each component to ensure that each, separately, be as near perfect as possible. (This leads to a wealth of "sacred" specifications that will be supported in the mind of creative man by the early "believe teacher" training. He will then reject any pressures to depart from his specifications.)[7]

McLean had fought each of these organizational stumbling blocks through his experience as inventor, facilitator, and manager of new weapons.

Develop before You Produce

McLean rejected the idea that the creative process could be planned. The worst aspect of such planning, he thought, was the creation of premature operational requirements for weapons systems. In discussing the differences between Falcon and Sidewinder, for instance, McLean emphasized that Falcon had a formal military requirement, while Sidewinder lacked one. Thus the Sidewinder project could create the best solution for the users in terms of what the laboratory was actually able to do.

In the abstract, the premise behind the operational requirement process made sense. If the services knew exactly what they wanted, their request should be as specific as possible. In reality, however, a requirement implied that those requesting knew what they could get—but in fact they couldn't know because no one knew what was possible until research had explored the possibilities.

Operational requirements—generally the result of intensive negotiations—asked for things that laboratories might or might not be able to produce. Whether they could be delivered was unknown until experimentation had taken place. Several problems followed from starting with a requirement:

A requirement might ask for something that was beyond the current state of the art. In this case, R&D time would be wasted trying to achieve the impossible. The result might well be a highly compromised system, impossible to abandon because of the bureaucratic system, but useless in warfare.

A requirement might fail to ask for something that was valuable and possible and could turn out to be a better answer to the operating units' problems. As with the Sidewinder, requirements might actually interfere with research on emerging solutions.

Valuable ideas might emerge from the laboratories and be refused because no one had requested them. This is what happened originally with the AR-15 rifle.[8]

The requirement process might thus make orphans of systems worthy of support while supporting those that needed termination.

What was needed, McLean thought, was to open a creative dialogue between operating units and laboratories. Instead of a linear process, McLean wanted to encourage more of an iterative feedback process.

As R&D proceeded, so would a dialogue with the users, to make sure that options the users chose were those that R&D could deliver—and the best that R&D could deliver. The key point, for McLean, was that the requirement not become fixed before the research was finished. Vice Adm. Bill Moran, thinking about this process, made a similar observation: "Now, you had that in Sidewinder. *The requirements people and the developers were both right here.* Nobody else was giving any requirements. We kept changing the requirements to fit the facts of life. You can do it this way, and then when we get better later on, we'll do it [another] way."[9]

Once one knew how the system ought to be designed, then it could be produced in quantity. The R&D part of the system, then, should be given only very general goals; the details of system specification should follow research, not precede it. China Lake tried to have both designer and user present during creation.

McLean was sensitive to the snowballing momentum of a weapons system once it entered the development process. It was dangerous, he thought, to commit a system to production before research showed that it would work. Careers of military officers became dependent on the system's acceptance and success. Money might keep pouring in even though the system failed to work. The result was often a financial quicksand and an operational nightmare.[10]

This premature commitment often jeopardized the necessary honesty of the inquiry process. Such commitment, McLean thought, should come only *after* research had shown a system to work. Specification in detail was important—but only at the production stage. Underlying McLean's strategy for R&D, then, was the conviction that detailed planning was possible only after a system had been shown to be both technically feasible and responsive to the user's need. Premature planning put a straitjacket on the creative process.

Essential to the ability to inquire freely, McLean thought, was to subject the weapon designer's intuitions to rapid translation into prototypes

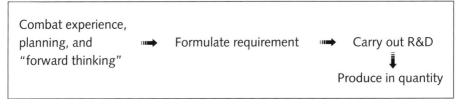

Standard American weapons systems acquisition process.

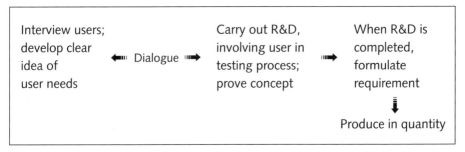

McLean's vision of the ideal R&D process.

and immediate tests. Design, test, and development had to be part of a single, unified process. A "full spectrum" laboratory like China Lake was essential because the feedback from experiment could be translated into new design ideas and more effective weapon systems.

McLean noted that many good ideas originated with the laboratories and then had to be sold to operating units. Previous studies have shown, for instance, that a working prototype was often far more convincing than mounds of paper studies.[11] This was often the model for China Lake. In R&D parlance, this is known as "science push" as opposed to the "demand pull." Reconnaissance expert Amrom Katz, an astute observer of the R&D system, claimed that in certain areas, virtually all useful military systems originate in this way.[12]

Yet as I discuss in chapter 18, the traditional system has survived all such intellectual challenges. While McLean's philosophy worked on Sidewinder, and other China Lake projects, the navy, far from studying and copying the China Lake model, eventually decided instead that it represented a deviation from the straight path. Its response to China Lake's deviance was to bring it back into line.

[14]

Early Generations

Sidewinder is the Smith missile.

Newt Ward

Falcon, Sidewinder, and Sparrow—the three air-to-air missiles that passed operational evaluation in the 1950s—seem to be with us today. Sidewinder and Sparrow still resemble the early models; Falcon has since mutated into the larger Phoenix. The early forms laid down the lines along which their successors evolved. Were the early forms the right ones? Or does the system have trouble developing a new missile from scratch?

If the outside forms have changed little, what is inside the missiles has changed dramatically. The latest models have better seekers, smarter fuzes, longer ranges, and deadlier warheads. Radio tubes have given way to microchip technology. Resistance to countermeasures has improved. Sidewinder is a far different missile today than it was in 1956, the year it joined the fleet.

Since 1964 U.S. air-to-air missiles have used the abbreviation AIM for "air intercept missile." Sidewinder 1 was the AIM-9A. Sidewinder variants subsequently were indicated by the letter following the numeral 9. The first Sidewinder to get this designation was the AIM-9B (Sidewinder 1a). It was B because it was the second Sidewinder. Smaller changes were signaled by modification numbers; there was AIM-9B Mod 1, Mod 2, and so on. The AIM-9B had fourteen Mods!

Since then, Sidewinder has almost gone through the alphabet. The AIM-9X is under development.

Limited Lease

With Sidewinder's success, the navy gave China Lake a wide license to innovate. Yet this license was only as good as China Lake's ability to fulfill the implicit contract that went with it. The contract implied that China Lake would remain a technical leader whose internal processes gave it a lead over industry. China Lake, then, had to maintain its creativity to keep its license. Industry, however, had no interest whatsoever in allowing China Lake to keep this license and tried to get it canceled.

[171]

Fortunately, until 1967, China Lake had McLean to protect it, and in the late 1960s a new generation of leaders emerged to continue this protection. For most of this time, Sidewinder was China Lake's flagship weapon, and China Lake oversaw the birth of several Sidewinder generations, each the product of a somewhat different team, although the generations overlapped: the technology of the latest version was being tested in the lab with parts from the current production model.

Sidewinder technology also resided in the labs of contractors such as Philco (Ford) and Raytheon, and in time they began designing their own prototypes. Many scientists and engineers left China Lake and joined contractors, trading the austere world of the desert for the higher salaries and urban comforts of industry.

When everything about guided missiles was brand new, Sidewinder was a skunk works project. Later, creating new guided missiles became a more routine process. Contractors competed for development contracts (big money) and for production contracts (bigger money). At any given time, several different concepts were under consideration.

Not surprisingly, industry often suggested that Sidewinder had run out of steam. It proved hard to replace, however. Just as the navy seemed ready to give up on it, another version came along. Since alternatives like Agile failed to materialize, both the navy and China Lake were reluctant to destroy a working system. Until the AIM-9R version, furthermore, China Lake seemed to be invincible.

The navy waited to see what China Lake would come up with; industry waited for China Lake to fumble. It was a long wait. Over four decades, several generations of Sidewinders have given pilots the edge:

The first generation included Sidewinder-1, 1A, and 1B. This generation had an uncooled lead sulfide (PbS) seeker, a tail-attack-only envelope, and limited range.

The navy's second generation included the AIM-9C and 9D Sidewinders, originally known as SARAH (*semiactive radar alternative head*) and IRAH (*infrared alternative head*). These Sidewinders had better seekers and longer ranges. This generation also included improved versions of the 9D: the AIM-9G and the AIM-9H. The air force ignored China Lake for its second generation and got Philco (Ford) to provide improved heads for some twelve thousand AIM-9Bs. These became the AIM-9E, AIM-9F, and AIM-9J.

The third generation began with a joint navy–air force missile, the AIM-9L Super-Sidewinder. It was followed shortly by the AIM-9M and AIM-9M(R), which incorporated counter-countermeasures. The 9M(R) refers to a retrofit of 9M including modifications 6, 7, 8, and 9.

The AIM-9R was an attempt at a major change in imaging, replacing the gyro with a charge-coupled device (CCD). It passed its tests but never reached operational status.

The AIM-9X is the latest generation. The new Sidewinder will have an imaging seeker, high gimbal angles, helmet-mounted sights, and vectored-thrust control. A "true dogfight missile" may be at hand.

Sidewinder 1

Sidewinder 1 is the missile whose development I have chronicled—
the missile that first passed all its tests and went to the fleet. Its half-mile range at sea level and very restricted envelope, roughly plus or minus 20 degrees off the tail of the target, left much to be desired, but it was still superior to an unguided rocket.

Its biggest technical problems were its ball gyro seeker, its short guidance time, and its lack of rollerons. The ball gyro caused it to corkscrew and was soon replaced by a more conventional Scotch-cross gimbal, which solved the problem. The short guidance time was fixed with a longer-burning gas grain.

Without rollerons, Sidewinder did not perform well against targets operating above forty-five thousand feet.[1] Potting the preamplifier to keep down vibration allowed the use of better rollerons to solve this problem. In 1956, even as the Sidewinder 1 became operational, changes were underway to bring out the improved version, the 1A.

Sidewinder 1A (AIM-9B)

The AIM-9B belongs in a special class as one of the great weapons of all time; it changed the nature of air warfare. Chuck Smith, a junior member of the Sidewinder 1 team, became the program manager. Afterward, until 1976, he filled a variety of important roles in designing, leading, and managing Sidewinder development efforts with the AIM-9D. He also served as midwife to the AIM-9L.

The new version had Don Steward's Scotch-cross gimbal and twice the guiding time of its predecessor.[2]

A January 1957 BuOrd evaluation tested the first thirty missiles off the Philco line; the missile then passed its operational evaluation and went to the fleet in July 1957. AIM-9B first saw combat off Formosa (see chap. 16). It was also the first Sidewinder used in Vietnam. On 12 June 1966 Cdr. Hal Marr in a navy F-8E downed a MiG-17 with an AIM-9B.[3]

Although it was a big improvement over the Sidewinder 1, and clearly superior to Falcon and Sparrow, the AIM-9B was an antibomber weapon first and a dogfight weapon second. Its probability of success under ideal conditions was about 70 percent, and it was constrained by a relatively narrow firing cone off the tail of its adversary. Its range was about one mile at sea level, although it could go about two and a half miles at 30,000 feet. It had rollerons for roll damping at high altitude, and would work, at least in principle, up to 80,000 feet.[4] However, at low altitudes, combat showed that the operating envelope shrank, as the ground interfered with IR detection. More than 95,000 AIM-9Bs were manufactured—40,000 by Aeronutronic (Philco-Ford); 40,000 by General Electric; and another 15,000 by the German firm Bodensee Geratechnik.[5] Following the support rendered to Taiwan in 1956, the United States began to arm its allies with the missiles. Friend and foe alike also produced copies—more or less faithful—of the missile. The Israelis improved on it with their Shafrir and Python.

The AIM-9B made McLean a cult hero in the former Soviet Union engineering community. While the comparable Soviet air-to-air missile had thirty-four radio tubes, Sidewinder had only fourteen—and only nine of these were used during flight, which gave rise to the oft-quoted "seven or eight tubes." When McLean received his presidential award in 1958, a group of missile engineers in Moscow drank a toast to him.[6]

The 9B went through fourteen mods, the last ones stopgaps to compensate for the slow development cycle (seven years) of the Sidewinder 1C.

Sidewinder 1B

This version was to have been carried internally and was designed especially for the Grumman F11F Tiger. To fit the Tiger, then undergoing final development, the 1B was to have folding canards and fins that would allow it to stow in a long rectangular space 106 inches x 7.75 inches x 7.75 inches. The concept might have worked, but in the end, the Tiger carried its Sidewinders on its wing pylons. The Avion Corporation, in Paramus, New Jersey, had done excellent work in designing the 1A guidance head and was selected to design the head for the 1B. Unfortunately, Avion hired a new team for the 1B work. It simply did not seem competent, and following a recommendation from engineer Ken Powers, Walt LaBerge—by then head of the Sidewinder project—decided to cancel it.[7]

The Albuquerque Tragedy

When a Sidewinder accidentally destroyed a manned B-52 bomber in 1961, killing three of its crew, one author described it as a "once in a lifetime coincidence." It was a coincidence, however, that could have been

[174]

prevented. The air force was practicing intercepts with live missiles and the validity of Murphy's Law—if anything can go wrong, it will—was demonstrated once again. The B-52 shoot-down was an accident that happened because of the convergence of several bad practices.[8]

The first error lay in not fixing a known problem. During the second Holloman shoot-off, Chuck Smith had found that a hydraulic line on the F-100 Super Sabre had crushed a Cannon plug. He rerouted the connection between cockpit and missile to avoid this problem, but his on-the-spot fix was neither incorporated in production aircraft nor retrofitted. The standard F-100 Super Sabres, when outfitted with Sidewinders, got the same faulty arrangement as the one that had failed in tests.[9]

The second error was an air force decision not to order dummy training rounds of Sidewinder, although the reason for the decision is not clear. Tierney had been shocked to find out from an air force civilian official that the air force did not feel it needed training rounds. Tierney felt that the civilians involved just did not understand their importance.[10]

[175]

The third error was that the Air National Guard apparently used missiles that never should have been mounted on planes. In a later investigation, Bud Sewell would discover several missiles in the Air National Guard inventory with expended gas grains, which meant they were unable to guide. While this did not cause the B-52 accident, it was an indication of careless maintenance procedures.

The scenario began unrolling on 7 April 1961, as the players assembled in the skies over New Mexico. Two Air National Guard F-100 Super Sabres with badly wired but live missiles were stalking a B-52.

Under aircraft commander Capt. Don Blodgett, the B-52 *Ciudad Juarez* took off early in the morning on a routine training mission. The bomber belonged to the 95th Bomb Wing of the Strategic Air Command stationed at Biggs Air Force Base, El Paso, Texas. The F-100s, circling at thirty-eight thousand feet awaiting the B-52, belonged to the New Mexico National Guard's 188th Fighter Interceptor Squadron. The lead plane was flown by 1st Lt. James Van Scyoc; Capt. Dale Dodd was on his wing.

The F-100 pilots verified that their 20-mm cannon and Sidewinder missile switches were in the "off" position—and repeated the armament safety check after the ground controller queried them about it. Confident that everything was okay, they maneuvered into position for a simulated gun pass.

The fighters roared past the bomber on the first simulated pass, then joined up and turned back toward the bomber for a second run. Van Scyoc was closing on the B-52 when he felt a slight jar and then saw one of his Sidewinder missiles streaking toward the bomber. He called out "Look out! One of my missiles has fired!" But it was too late.

The missile struck and the giant bomber rolled into a tight turn to the left. The high G-forces made escape difficult; not all inside cleared the aircraft. The F-100 pilots called for help while looking for parachutes. When help arrived, three dead crew members were found in the wreckage; five other crew members, several seriously injured, were rescued.

Early in the investigation, it became clear that pilot error was unlikely to be a good explanation. A Department of Defense investigation team, including China Lake's Bud Sewell, found that the wing Cannon plug had been damaged, apparently by the same hydraulic line pounding that Smith had uncovered during the second Holloman shoot-off in 1956. Moisture had penetrated into the plug and frozen as the F-100s climbed up through clouds. On the way down, the moisture melted and shorted out the firing circuits to the missile gas generator and rocket motor.

Because Smith's fix had not been *standardized,* all the F-100s configured to fire Sidewinder had the same faulty wiring arrangement. Further investigation showed that almost all the F-100s in the National Guard squadron had the same damaged Cannon plugs.[11]

The Second Generation

Sidewinder 1C (the AIM-9D)

The Sidewinder 1C began the second generation of expanded-envelope Sidewinders. An expanded envelope was needed because, for instance, it was possible to escape the earlier AIM-9B by turning tightly—if a pilot saw it coming.[12]

The original Sidewinder 1 team had scattered, leaving Frank Cartwright, Smith, and Benton in the Weapons Department. They became the nucleus for the development of the Sidewinder 1C. In addition, China Lake decided to develop not only an infrared head [176] but also an alternative *radar* head for the missile—the infrared alternative head (IRAH) and the semiactive radar alternative head (SARAH). The IRAH became Sidewinder 1C Mod 29 and the SARAH Sidewinder 1C Mod 30; the missiles were identical except for the seekers. Trademark problems soon caused the team to drop the SARAH acronym, however. The Weapons Department developed the infrared version, while Tom Amlie, still in the Aviation Ordnance Department, developed the radar version. Production of the older AIM-9B was being managed out of the Engineering Department.

Smith was responsible for the update of Sidewinder missile technology from the Sidewinder 1A to the 1C (that is, AIM-9B to AIM-9D), and his team made the AIM-9D a superior missile. The gyro optical system was completely redesigned; Mickie Benton's optics branch did this work with mechanical engineer Eddie Allen playing a key role, as he did on several free-gyro seeker designs. The 9D's seeker used high-pressure nitrogen gas cooling to increase the sensitivity of the lead sulfide cell. The infrared (AIM-9D) version originally was designed to use a lead selenide detector. Problems in making these detectors at Santa Barbara Research Center led China Lake to develop the cryogenically cooled lead sulfide detector instead.[13] Since the missile was relatively small and still had bulky tube electronics, it did not have room for an internal cooling gas supply system. Instead, a high-pressure nitrogen bottle was fitted onto the launcher.

The seeker's Joule-Thomson cooler could run for four hours using this bottle. With the detector cooled as low as a frigid 77 degrees Kelvin, sensitivity was much higher, and a longer wavelength spectral band could be used. The seeker could now peer through haze far better than had the 9B. The cooler was expensive, but by 1962 it was evident that the 9D was going to be a lot better than the 9B. It was far superior in gaining and maintaining a lock on targets. Equally important was the AIM-9D's more powerful rocket motor and larger control surfaces.[14] By using a smaller (1.8-inch) primary mirror in the seeker, the 9D got a sleeker profile with less drag; its theoretical range at altitude was eleven miles, triple the 9B's range.

Philco-Ford was chosen to work with China Lake on developing the new missile, but there was much friction. Ken Powers, working with Smith, commented that carelessness and neglect were typical of Philco's approach. Prototypes were not built according to drawings; test instruments were not properly calibrated; and Philco reports showed that many missiles had failed tests repeatedly.[15] As Ken Powers said:

> I looked at their internal records; one missile had gotten into final test thirty times, and every time had failed because of reliability prob-

[177]

lems, cold solder joints, burned-out resistors, open connection on the circuit boards, etc. I asked Bill Woode, the NOTS main contact at Philco, "How can you possibly expect us to use this once we get it, because obviously it's very unreliable?" He said, "That's not my problem. I'm only trying to get it through final test. I don't care what you do with it when you get it." Now with a team member like that you don't need enemies.[16]

These circuit board problems delayed flight tests for two years while the problems were ironed out.

Testing of the AIM-9D for the fleet involved some colorful episodes. Tom Amlie described one that occurred in 1964 during the operational evaluation briefing at Point Mugu. "I emphasized several times that they must visually identify the target before firing." Noting that one of the pilots was a marine captain, Amlie made his presentation blunt and direct.

At the end of the day, I came back to the ready room and there was my captain, his face almost as green as his uniform. He said, "Doctor, can I buy you a drink?" I responded that anyone could buy me a drink. We went over to the little bar in the Bachelor Officers Quarters, got a couple of cold ones, and he told me his story. He had had a good radar lock and a strong growl from the missile. He was about to fire and then remembered how I had stressed the visual ID. He popped afterburner for a few seconds and found that he was locked onto a TWA 707 letting down into Los Angeles International Airport. My response was to pop for two more beers.[17]

Point Mugu had had a lot of trouble with airliners cutting through the warning area—but a letter describing this incident put an end to it.

Production Saga

McLean had designed Sidewinder for easy production, and China Lake stayed aloof from the production process until the 1960s. Then problems surfaced with the AIM-9B at Philco and the AIM-9D at Raytheon. Production engineer Burrell Hays was encouraged to fill the gap and slowly found his niche as the apostle of production quality. He knew that the key to good production was a proper data package that would permit the government to get competitive bids on sub-systems or even to develop a second source.[18]

To get a feel for Sidewinder production, he visited Philco's 9B production line. (Philco was producing 9Bs in addition to developing the AIM-9D. It intended to bid for the 9D production contract.) Hays and fellow engineer Glenn Hollar were stunned by what they saw.

[179]

Philco was building the 9Bs at the old, dilapidated Atwater Kent Radio Company plant on Wissahickon Avenue. Built into the side of a hill, the plant stood about eight stories high. The missile parts came in on the bottom floor, then endured a tortuous journey up and down between the floors as parts became subassemblies, which in turn became assemblies. Parts as well as product quality got lost in these journeys up and down. Problems were made worse by mobile partitions that separated various activities and turned the floors into mazes.[19]

Philco was supposed to be building Sidewinders in a clean room, but the factory was filthy. (Hays later found similar problems at Raytheon's Lowell, Massachusetts, plant.) Hays felt that Philco's "zero defects" program was a joke; only 23 percent of the missiles that came off the assembly line worked. Paperwork on the missile was hard to find. "It was obvious even to a novice that they weren't doing the same thing twice."

Hays decided that quality had to be built into the production contract, which was vague and left Philco far too much freedom. For answers, he consulted the government inspectors. The inspectors turned out to be true professionals, and they took time out to teach Hays the way the job ought to be done. Previously, they had gotten little support from the navy.

In searching for appropriate models, Hays also discovered that the National Aeronautics and Space Administration and the Burroughs Corporation had done considerable work on reliability. "Plagiarizing heavily from their manuals," bit by bit, he put the necessary changes in place, and the contract was rewritten in more exacting terms.

For instance, a solder specification was imposed on the five thousand soldered connections in the missile. The Naval Air Systems Command (NavAir—formerly BuWeps) was nervous about this, but Hays was persistent. Teamwork between engineers and technicians was imposed on Philco. The contractor protested, but Hays was determined. He started demanding quality on the AIM-9D from his old friend Walt LaBerge, who had moved to Philco from China Lake.

Hays found an ally in John Rexroth, the senior civilian design and production expert at NavAir. Rexroth imposed a quality assurance specification that mandated controlled factory conditions as well as tighter product specifications at Philco. Even without design changes, these improvements in process increased the yield of useful missiles coming off the line from 23 percent to 87 percent.[20]

Hays also advocated second sourcing. NOTS suggested to the Bureau of Ordnance that other firms be invited to compete to build Sidewinders. The navy found that interest was strong, and when the first phase in the competition was over, eight firms had become finalists. Sidewinder had barely begun production at Philco before the navy brought on board General Electric's Light Military Electronics Department in Utica, New York, as a second source for the Sidewinder guidance-and-control unit. Ironically, however, the first missiles from General Electric tested badly. GE agreed that the first missiles off the production line would be tested, but when the missiles actually arrived at China Lake, they did not pass even the bench tests. One missile was missing a detector cell; another had tobacco flakes in the dome. There was a big management flap about this. Strangely, enough, however, when the missiles were actually fired, eight out of ten worked. It was suspected that General Electric had "swapped the rounds at midnight" for ten in better condition. It was subsequently rumored that the General Electric representative was fired over these events.[21]

Starting with the AIM-9B production improvements, Hays and his colleagues went on to improve production not only of Sidewinder, but also Shrike, Harpoon, Condor, Sparrow III, and even Phoenix missiles. China Lake evolved a remarkable quality control program that became a national standard. Hays also constantly sought second sources for China Lake missiles.[22]

[180]

Raytheon Starts the Learning Curve
In 1964, as China Lake and Philco-Ford wound up the AIM-9D development, the navy asked for production bids on the first lot of guidance-and-control units. Philco-Ford, longtime 9B producer, and newcomer

Raytheon bid for the major contract to produce 4,450 units. Philco-Ford offered to build guidance-and-control units at about $8,000 per copy; Raytheon's bid was $3,500.

China Lake, estimating that $6,000 per copy might be a reasonable price, figured that Raytheon was trying to "buy in." Smith told navy procurement officials that Raytheon probably would take a financial beating if its bid were accepted but was told: "They're big boys; after all, they build the Sparrow!" Nonetheless, the navy, without explaining why, offered Raytheon a chance to rebid. Thinking they were being asked to *lower* their price, Raytheon left the bid unchanged and the navy accepted.

[181]

Raytheon had been manufacturing Sparrow IIIs for years, but Sidewinder required new skills. By 1966, four to five months into its production of Sidewinder, the company was in trouble. During development with Philco, circuit board problems had delayed flight tests for two years.[23] Now, because development had taken so long, Raytheon and the navy had decided to skip a pilot-production line and go directly into full-scale production.[24] The result was disaster. Raytheon had eliminated design engineers on the project and used only production engineers. Prototype drawings supplied by China Lake, however, had serious problems with dimensions and tolerances.[25]

In January 1967 the production line was shut down, and the China Lakers sat down alongside the Raytheon production and design engineers to fix the problems. When the improvement program began, Raytheon countered with a lawsuit, but in the end the company had to comply. Up to this point Raytheon had been running three shifts of twenty-one people, trying to produce three hundred missiles a month, but actually delivering only one hundred that met specifications. On the road to ruin when the China Lakers arrived, Raytheon quickly got the system operating and was able to produce three hundred missiles a month with only eighteen people per shift.[26]

How many of these problems were Raytheon's fault and how many were China Lake's is not clear. China Lake had supplied a less-than-adequate data package. In any case Smith decided that Raytheon needed to make twelve major changes in the way the missile was produced. In the interests of getting Sidewinder to the fleet quickly, he approved all twelve changes while he was in Massachusetts. Burrell

Hays became angry when Smith indicated that he had OK'd changes costing millions of dollars. In fact, Raytheon sued over the required changes and got the navy to pay an additional $14 million. In the end, the contract cost the navy about what Philco-Ford had originally bid.[27]

In time, China Lake and Raytheon resolved their differences. The AIM-9D proved to be a fine missile that was used extensively in Vietnam, and Raytheon began a long-term relationship with Sidewinder and China Lake.[28]

Raywinder

Even before problems with Sidewinder 1 were ironed out, radar-detecting Sidewinders were being worked up in the laboratory. John Boyle was working on a version of Sidewinder—Raywinder—that would home on radar antennas.[29]

By May 1956 Boeing was interested in arming its bombers with Raywinders for use against aircraft jamming their radars. Seekers were tested from the air against ground radars beginning on 26 October 1956; they had large miss distances initially but got better.

In May 1962 a Raywinder missed its radar target by nine inches, earning a "well done" from Admiral Withington at BuOrd. But Raywinder's original purpose had been to hit radars carried by Soviet fighters. The problem was the Soviets were not putting radars in their fighters. And ground radars could be attacked with Shrike, a bigger missile with a larger warhead and longer range that was under development. The Raywinder project was closed; Boyle moved on to other things.[30]

SARAH (AIM-9C)

By 1957 Amlie was working on SARAH, the "semiactive radar alternative head" version. This missile would bring the long-sought all-weather capability.

A *radar* Sidewinder? Absolutely! McLean believed in flexibility and wanted a radar version because the AIM-9B was limited to tail attacks. SARAH could provide a fleet-defense weapon for the smaller *Essex*-class carriers, whose Crusader aircraft could not carry Sparrows.[31] The project started slowly but soon became a full-intensity one. In August 1957 Amlie left the infrared test program and started directing work on SARAH.

One of the first challenges was to develop an effective guidance unit, which Technician Earl Donaldson did under Amlie's direction. Don Steward, who had designed the gimbal for the 9B, also did design work for SARAH. The seeker became an elegant mechanism that worked well.[32]

In the meantime, glitches claimed the first two air firings, and the project was nearly canceled. Getting political support in Washington was a constant challenge. Liz Beggs at NavAir was the AIM-9C sponsor. "She was determined to be the best missile person in BuWeps, and always raised hell with me about the 9C," Amlie recalled. A successful third shot in October 1959 saved the program. The AIM-9C avoided the "near-miss" problem that plagued the Falcon because Amlie took time to read the Falcon playbook. When the Falcon engineers found that Amlie had fixed the near-miss problem, they wanted to know what he had done. "I read your technical reports," he explained. Hughes had outstanding technical reports and Amlie read them carefully.[33]

[183]

Getting an adequate radar into the Crusader was a problem. In May 1960 Amlie and Cartwright were called to Washington to defend data they had submitted on AIM-9C performance. Their numbers were based on a twenty-four-inch-diameter radar, but the navy had specified a smaller thirteen-inch dish that was far less effective. The Office of the Chief of Naval Operations had gotten the numbers mixed up and wanted China Lake to justify its apparently exaggerated claims. Tom Amlie recorded his impressions shortly after the trip:

There were about 30 people there including eight captains. I sneaked in a blackboard and went through some of the radar range equations. There were 15 minutes allowed for questions and a rather lively discussion took place. After about 30 minutes of questions, Mr. Caspar [a civilian] stood up and said that time had run out and we would have to quit. I got up and said that I had come 3,000 miles to answer their questions and was going to do it. Several of the captains told Mr. Caspar to sit down and be quiet.

The discussion became quite spirited and Frank Cartwright, who felt obliged to defend his honor, called upon the deity a couple of times. I confess that I addressed this august group as "You clowns," several times. It developed that what was hurting them was that they had an excellent airplane and probably an excellent missile, but they had a little peanut-sized radar with a 13-inch dish. I pointed out that I had been there a year-and-a-half before with tears rolling down my honest cheeks making a strong pitch for a larger dish to anyone who would listen. They freely admitted this. The best statement of the situation came from one of the captains who said, "But God-damn it, Doctor, you should have made us put in a larger dish!"[34]

In the end, a larger radar dish, the Magnavox AN/APQ-83, was installed on production versions of the plane.

While fleet introduction was originally planned for the summer of 1961, SARAH actually went through operational evaluation in 1964, the same time as IRAH (the AIM-9D). During the operational evaluation of the SARAH, the missile showed a 77 percent single-shot kill probability (SSKP), not bad for a weapon just out of the development process. Both versions of the AIM-1C did so well that the China Lake team found itself in demand by industry. Raytheon offered Amlie a lucrative job "to come to Oxnard and fix the Sparrow's problems," but he declined. Raytheon also offered Smith a job; he, too, declined. Smith eventually did join Raytheon twelve years later, when Rear Adm. Rowland Freeman had assumed command at China Lake. The Raytheon job suddenly began to look attractive.[35]

SARAH officially became the AIM-9C and IRAH became the AIM-9D. Two NAS Miramar-based F-8 squadrons deployed with the missiles in the fall of 1964, and China Lake's role shifted from development to technical support and pilot training, a job it continued to discharge until the navy developed special training squadrons.[36]

Hands-On Assistance

[184]

Amlie himself went on board three *Essex*-class carriers for the AIM-9C's introduction. Having a Ph.D. leading the program onboard ship sent the unmistakable message that China Lake stood behind its weapons. The radar missile, though accurate, required more care and feeding than its infrared brother. The biggest problem, however, was not in the missile itself but in the Crusader's radar, which failed frequently and was difficult to repair.[37]

Amlie went about solving the problems in two ways. First, he made sure the mechanics had enough tools to do the job. Tools were a chronic problem; getting them required signing considerable paperwork—and they tended to disappear, which meant the mechanic who signed for a particular tool had to pay for it. Amlie gathered up bundles of tools at China Lake and presented them to the mechanics with no strings attached. "No paperwork. No nothing. I owned those kids," he said.   [185]

Amlie's second fix concerned the radar's twenty-thousand-volt power supply. The high-voltage power supply that ran the cockpit's cathode ray display tended to short-circuit at altitude. The radar itself was pressurized in flight but it was connected to the display with one of those connectors (unpressurized) that McLean so despised. The connector was the source of the shorts. Amlie's solution was to use insulating grease to fill the holes in the ceramic insulators holding the radar's wires.

> Every time we had a short and pulled it apart, there was a carbon track on the connector's sleeve. I took to carrying a penknife with a thin, sharp blade to scrape off the carbon track. I would then fill the sleeve with Dow-Corning 200,000 Centistoke silicone grease. I also began giving the leading maintenance chiefs a knife like the one I used. I sure didn't make any enemies there. This short was about the only real problem in the Magnavox radar—which was about ten times more reliable (in terms of mean time between failures) than the disastrous Westinghouse radar in the F-4.[38]

After all this assistance to the mechanics, the China Lakers could do no wrong.

Many USS *Hancock* (CVA-19) pilots doubted that the new radar Sidewinder would work. They associated the Sidewinder with a tail shot—which indeed was necessary for infrared homing—and doubted that the AIM-9C actually could tackle another fighter head-on. Amlie collected drinks from the pilots by betting the missile would work.

Crusaders with AIM-9Cs provided the *Essex*-class carriers with a head-on attack capability, and the missile was mated to radars on the F-8C, D, and J versions of the Crusader. It thus became part of an integrated "weapons system," which, of course, is just what killed it. The Crusaders were aging by the mid-1960s, even though Crusader pilots shot down more than their share of MiGs during the Vietnam War. As long as smaller carriers remained in the fleet, so did the Crusaders with their radars and their AIM-9Cs. When the smaller carriers were replaced by larger ones with F-4 Phantom IIs, the Crusaders and their AIM-9Cs got retired together. The very feature of the AIM-9B Sidewinder that made it so popular—that you could hang it on nearly anything—assured it a long life. The AIM-9C, however, lived and died with the planes using compatible radars. Later, the AIM-9C would have a second life as Sidearm, a small antiradar missile, mounted on navy helicopters and fixed-wing aircraft.

The Air Force Models

AIM-9E and AIM-9J

By the end of 1956 the air force had accepted Sidewinder 1s for operational use; by the fall of 1957, air force units began to get the better AIM-9B Sidewinders. Chuck Yeager's squadron of F-100 Super Sabres was among the first Tactical Air Command units to get AIM-9Bs:

> TAC gave us the first Sidewinders, eager to discover how quickly their best pilots could become proficient learning to fire weapons that cost $15,000 apiece. At those prices we didn't waste many practice shots. Firing those Sidewinders really impressed us about how well we would need to fly to survive in future combat. All we had to do was wax a tail, turn on the system, and get a rattling tone in our headsets, which meant that the heat sensors were locked on the hot air exhaust of another jet, turn on our gunsight radar that locked us on target, then fire and watch that missile streak right up the tail of a drone, blasting it to pieces. Until evasion tactics could be developed, the price of getting your fanny waxed in future combat would be a high-explosive missile rammed up your behind.[39]

[186]

The pilots liked the new missiles, but the air force leadership was not thrilled about accepting a navy missile and only grudgingly kept accepting navy Sidewinders into the 1960s. When the AIM-9D came along, the air force decided it wanted no part of it and opted to design its own Sidewinders. So the air force got Philco-Ford to design versions of Sidewinder to air force specifications. This design effort did not work out well.

To create what became the AIM-9E, Philco-Ford installed new heads and control surfaces on the 9B frame and motor. In principle, the air force specifications were crafted to produce a true "dogfight" AIM-9B, with a wider gimbal angle and reduced-drag guidance system enabling the missile to turn tighter. Unfortunately, the new missiles lacked the range and the discrimination of the AIM-9D, which may explain a puzzling finding on kills in Vietnam. While most early air force air-to-air kills were made with Sidewinder, the Sparrow III later became the most common air force MiG killer. The pilots may have become disenchanted with the 9E's limited range and performance, and now had better operational intelligence, which allowed firing Sparrows beyond visual range.[40]

The air force AIM-9J was a solid-state version of the 9E with superior control surfaces. Nonetheless the air force Sidewinders were poor substitutes for the navy versions.[41] Insiders considered both these missiles dogs in relation to the navy AIM-9Cs and AIM-9Ds, and some air force pilots wanted to put navy launchers and missiles on their aircraft. Col. Robin Olds, Commanding Officer of Udorn Air Force Base, did just that, but had to use AIM-9Bs. Finally with the 9G, even the Seventh Air Force swallowed its pride and asked for the AIM-9Gs![42] But navy AIM-9Gs were not compatible with air force AIM-9E launchers. (The 9D's cooled detector required a liquid gas bottle, which the air force's 9E launcher lacked.) No doubt many air force pilots were relieved when both services were later forced to accept the AIM-9L version.

AIM-9G and AIM-9H

The AIM-9G was essentially a 9D with one key improvement: the Sidewinder expanded acquisition mode (SEAM), an idea that Smith's team had wanted to put into the original AIM-9D. Slaved to the acquisition radar, the seeker could revolve twenty-five degrees in a circular scan, instead of being aimed straight ahead, which improved the odds of acquiring targets. The 9G also featured a number of solid-state modules. Only about 1,850 of an order of 5,000 9D seekers were produced before the rest were modified to the 9G configuration.

As the AIM-9G program wound down, many at China Lake thought Sidewinder had gotten as good as it could get and needed only minor tweaking rather than radical innovation. As a result, the project was moved over to the Engineering Department. The prevailing opinion was that the AIM-9G was the last Sidewinder.

In December 1965, however, LaBerge (still working for Philco-Ford) reappeared at China Lake. McLean had some concerns about AIM-9G reliability and had asked LaBerge to consider options for change. Philco-Ford proposed converting Sidewinder to all solid-state electronics, circuit by circuit, over a period of time. The air force liked the idea, but engineer Walt Freitag did not. He thought it made more sense to convert the missile to solid-state all at once. He was impressed with the new integrated circuits and thought they would improve reliability. He convinced the navy that an all-transistor missile—which turned out to be the AIM-9H—was needed, and his branch then reengineered the missile with solid-state components throughout. While design specifications called for performance equal to that of the AIM-9G, the solid-state electronics in fact made the new 9H more reliable.[43]

China Lake developed the 9H in conjunction with General Dynamics, Pomona, which had recently finished a highly producible design for the Redeye shoulder-fired, IR-guided ground-to-air missile. Freitag's team used this expertise to design a package for components using a flexible Kevlar harness and "cordwood" modules.[44]

The project got stalled for six months, but politicking by China Lake and NavAir finally got it into production. The decision to go to solid-state technology was critical for Sidewinder's survival. It insured—as Norm Woodall put it—that "Sidewinder would be in the fleet forever." Woodall worked on a circuit that gave the missile a lead bias, aiming the missile forward of the heat source and solving a critical problem. When a target was in afterburner, Sidewinder would go for the "diamonds" in the plume instead of the airplane, which helped create the belief that Sidewinder could not hit an afterburner target. Lead bias got the missile to aim forward of the plume. Woodall estimated that thirty rounds were made and some twenty shots took place with the lead bias circuit in. This retrofit circuit was eventually killed, however, because the superior AIM-9L later completely replaced the AIM-9H. After a very successful firing program, AIM-9H entered fleet service in 1972—too late for the Vietnam War.[45]

Ground-Launch Versions

It did not take great imagination, given Sidewinder's versatility, to think of firing it from the ground. Indeed, China Lake had considered a ground-launch version as early as 1953. In 1958 BuOrd ordered a feasibility study for a weapon called "Hamburger" (the ground round). Cartwright took the lead in this study and analysis showing that Sidewinder could launch from a stationary launcher and guide toward a target flying overhead.[46]

Similar thinking led to a version for navy ships too small to handle the Terrier or Tartar missiles: the Osprey. Hamburger and Osprey got as far as initial tests but no further. The navy, busy trying to develop the close-in weapon system (which went to the fleet as the Phalanx 20-mm Gatling gun) was not interested. Smith thought that any competition for the close-in weapon system would probably be sent to Davy Jones's locker forthwith. This may be what happened to Osprey.[47]

The army felt that a ground-launch Sidewinder would be useful, so it began a development project with Aeronutronic (the former Philco-Ford), using the AIM-9D seeker. Under project manager John Lamb, Conrad Neal's group developed this system as Chaparral (MIM-72).[48]

The system was deployed extensively by U.S. forces. The Israelis also used Chaparral and shot down enemy planes with it in the 1973 Yom Kippur War. The first Chaparral kill was a Syrian MiG-17 in October 1973.[49]

An Era Begins to Close

Sidewinder was now twenty years old and its place in military history was assured. Going into its third decade, though, Sidewinder struck many naval officers as a weapon on its way out.

China Lake stayed vital into the 1970s, but things began changing in 1974 when Rear Adm. Rowland "Doc" Freeman arrived as commander. Convinced that China Lake's culture needed controls, he dictated major changes. He did not change China Lake completely, but his efforts did much to reshape the center's internal processes. Many felt his legacy was a weakened Naval Weapons Center. While there were later revivals of the proud and confident organization that China Lake had been, it never again rose as high.

[15]

Later Generations

My personal opinion is that we will have a Sidewinder-type missile in the Fleet as long as we have manned aircraft.

Frank Wentink, 1970

When production on the 9H was winding down, we just thought that we had transistorized the thing, we never thought there would be tens of thousands more missiles based on that design.

Dick Schmitt, Missile Designer

China Lake made immense contributions to the air war in Vietnam. It developed most of the new air-to-ground weapons used by U.S. forces and its Sidewinders performed well. But if pilots liked the Sidewinder, not many bureaucrats did. Nor did they like China Lake. Many felt both Sidewinder and China Lake had had their day; now it was time to move on.

And the United States did move on to develop other air-to-air missiles, such as the Phoenix and the AMRAAM (advanced medium-range air-to-air missile). The Phoenix, externally similar to Falcon, had an active radar seeker and a range of 120 miles. It was designed as a fleet-defense missile to be fired by F-14s fighting the outer air battle. Its cost and complexity are an order of magnitude higher than Sidewinder. The fleet got the Phoenix in 1973; it has yet to be fired in combat. The AIM-120 AMRAAM, also developed and deployed, is a good but expensive medium-range missile. These missiles, however, did not replace Sidewinder. In the Gulf War it was Sidewinder and Sparrow that got the air-to-air victories.

AIM-9L: The Super-Sidewinder

The birth of the third-generation Sidewinder was shaped by air-to-air missile problems in the Vietnam War, in which the navy and air force had used different versions and learned different lessons. The navy used good Sidewinders—the D, G, and H versions—and saw that many of the failures occurred because pilots had fired them without proper training or understanding. In conjunction with a general overhaul of air combat maneuver training—generated by the Ault Report (see chap. 16)—increased emphasis was placed on Sidewinder use at the new Top Gun Fighter Weapons School at NAS Miramar. This seemed to solve the navy's problems.

[191]

The air force, meanwhile, had been firing the E and J versions, which were *not* adequate. When the air force analyzed failures with *its* Sidewinders, the problem appeared to be reliability. The air force solution was to build smaller and cheaper missiles, and fire more of them, so the service approached Hughes, Ford, and General Dynamics to help. The air force was well into this program (known as CLAW) when William Perry, the Defense Department's deputy director of research and engineering (later secretary of defense) took charge. Perry told the air force to go instead to China Lake. The Pentagon had decided it was time for the two services to get together on the development of a common missile.[1]

China Lake welcomed the air force and treated the officers to a discussion of new technologies: wide-angle seekers, helmet-mounted sights, and thrust-vector control (steering jets instead of canards). Thrust-vector control promised missiles that could turn on a dime. Helmet-mounted sights—already tested in helicopters for miniguns—would allow faster pilot response. China Lake was putting together an advanced missile called Agile that would have all these features, as well as an advanced Sidewinder designated AIM-9K that would use only the wider gimbal-angle seeker.

China Lake argued that the new technologies would create a genuine dogfight missile. While the air force was familiar with these technologies—and was working on the AIM-82 missile that incorporated some of them—in the end it decided that the exotic technologies were too big a risk and opted for an upgrade rather than a new missile. Again, China Lake had just the program: the AIM-9H Product Improvement Package (PIP).

In 1970 Walt Freitag had been monitoring AIM-9H as it went into pilot production, assuming that it would be "the last Sidewinder." But Roland Baker, manager of the Chaparral program, decided that Chaparral could be improved and asked Freitag to carry out an "Advanced Chaparral Study." Freitag asked a young engineer named Dick Schmitt to help him and the two—recognizing that Sidewinder still had a lot of life—widened the focus to the entire Sidewinder program. The "Advanced Chaparral Study" soon became a "30-day study" to define what might be done with Sidewinder. The strawman missile that emerged was the AIM-9H PIP.

Raytheon, which had been making Sidewinders for nearly a decade, immediately went on alert. Ed Paul, Raytheon's Sidewinder program manager, had started a multiyear study of detectors in 1968 because he felt Raytheon had to keep working in phase with China Lake. When China Lake did its 30-day study, Raytheon ran its own 30-day study. When China Lake did a 60-day study following the 30-day study, so did Raytheon.[2]

China Lake decided the key to success was a better detector. Previous Sidewinders had used detectors that were best at detecting short- and medium-wavelength infrared radiation—the kind given off by jet engines and their exhaust plumes. But an indium antimonide detector would detect the longer wavelengths associated with warm aircraft parts, such as nose cones, giving the missile "all-aspect capability" (ALASCA), including the long-desired head-on shot. Radar missiles had this all-aspect capability, of course, but at the cost of higher complexity. Sidewinder kept it simple.[3]

Using indium antimonide for the detector was not a new idea. In fact, Benton's Optical Design Branch had wanted to put one into the AIM-9H, but the navy had shown scant interest in major changes to the Sidewinder, and in most respects the 9H was simply a solid-state version of the 9D. The next missile would be a chance to see what indium antimonide might do. Ed Paul immediately told his people at Raytheon to begin studying indium antimonide detectors.

Meanwhile, Ford Aerospace had approached Luke Biberman, then at the Institute for Defense Analysis, regarding Chaparral improvements. The navy then also approached Biberman regarding Sidewinder improvements. Raytheon decided it would support the navy's effort.

This led to a scramble between Ford and Raytheon over who would get the next Sidewinder development contract. Complicating matters, the navy wanted amplitude-modulated (AM) scanners while the air force wanted to use frequency-modulation (FM). Raytheon's Dick Beckerleg came up with a synthesis, an AM-FM system; this kind of problem-solving convinced Biberman that Raytheon was better prepared than Ford, and Raytheon got the job.

[193]

When it was clear that the air force was interested in the new detector, Benton's team got to work immediately on a feasibility study. Benton got Freitag's electronics group to make an absolute minimum of needed changes in signal processing and an indium antimonide detector was soon mounted on the missile. The modified AIM-9D was then taken out and fired from a Chaparral (ground) mount, the fastest way to get a test.[4]

For no apparent reason, the missile missed the drone by ten to fifteen feet. Nonetheless, this was enough to show the idea would work, and it gained the team approval for a development program. China Lake, Raytheon, and Biberman worked out final specifications, which were issued by Secretary of Defense David Packard in June 1971.

Raytheon, meanwhile, had developed the HX prototype—a Sidewinder with "rate bias." Like lead bias, rate bias aimed the missile forward of the hot tail or the plume in the afterburner. It was such a good idea that the navy forced China Lake to accept it. Getting it to work, however, proved difficult. In March 1972 Freitag's team carried out its first flight test at China Lake. The missile literally hit the drone on the nose. The team was elated, but the result was not what they expected: Why did it hit the nose?[5]

Yet the missile had worked. When many of the following shots passed in front of the drone instead of hitting it, however, the team knew it had a problem. Fortunately, the fifteen engineers were an active team under Freitag's gifted leadership. "It was a skunk works, a small number of people really making technical decisions," Schmitt said. Fortunately, also, General Dynamics had discovered and solved a similar problem with rate bias on its Stinger antiaircraft missile. Adding an automatic gain control to the rate bias circuit solved the problem.[6]

Changes were made quickly by the co-located design team. In his office—originally a library vault—Freitag held weekly design reviews with the aid of a blackboard. Just about the time the designers were getting smug about their concepts, he would ask, "Have you thought about . . . ?" and, typically, they hadn't. These design reviews were credited for many of the key breakthroughs. Even after promotion to division head Freitag still exercised leadership of the design team.[7]

The team was coached to insure that each circuit designer saw himself as a system designer, carefully considering the implications of each decision.[8] Using a small number of empowered people was the key to the missile's development. Elegant design was the objective. Schmitt argued for days with Raytheon's Dick Beckerleg and Ben Klaus about the right way to design a circuit, but whoever came up with one less resistor in the design won the argument.[9]

Allen Gates took over program management of the Sidewinder development team from Glenn Hollar. Gates had good technical credentials (in contrast with some program managers) and ran the program with a stern hand. But he also provided key technical input for the missile, especially the system dynamics of the airframe. It was Gates who, late in the program, decided to avoid too much new technology development and got the project focused on the key issues needed to get a producible missile. Previously, as an analyst, Gates had suggested taking some major technical risks. Once he became head of the program, however, he decided these options were too risky and used more tried-and-true technology.[10]

The air force was ambivalent during the design phase, and when troubles came, it was eager to dump the project. In 1973, when the 9L was in trouble, the air force wanted a shoot-off to show that its AIM-9J was not such a bad weapon after all. Raytheon brought a version of the AIM-9H to the tests, fitted with an internal gas bottle, so it would fit on air force launchers. The Raytheon "H Squared" proved far superior to the 9J, and the air force backed off.[11]

What would become the Sidewinder AIM-9L soon got back on track, but the air force, still skeptical of its ability to hit a plane in afterburner, scheduled fifty tests with QF-4 Phantom drones to verify the capability. The 9Ls took out one expensive drone after another. The air force showed up at China Lake, asking for a slight miss bias in the missile, so its drones would be spared. It got little sympathy.

While the engineers were fine-tuning the missile, it had to be sold to potential users. Ernie Cozzens, Freitag, Benton, and Gene Younkin went on the road. Shows were given to OpNav, NavAir, and even one to the air force at Wright-Patterson Air Force Base. NavAir's John Rexroth assisted the China Lakers; when necessary, he was perfectly capable of telling admirals to "sit down and shut up."[12] The team even got Hollywood to assist. One promotional film showed a captive missile with the new Chirp feature tracking a target. As the target went into a cloud, the seeker in the missile kept right on chirping, showing that the target was being tracked even in the cloud. All the efforts paid off, and the air force finally relented; air force captain Mike Hall, a Sidewinder test pilot, played a key role.[13]

Further politics intervened. A European weapons consortium headed by Bodensee Geratechnik—which had produced thousands of AIM-9Bs—had developed a seeker head for the proposed Viper missile that also would have an all-aspect capability. To bring BGT into the fold, the navy and Raytheon waived license and R&D recoupment fees, so that Britain, Norway, and Germany could coproduce AIM-9Ls to equip their own air forces. This meant the end of the Viper. In the end, Japan also produced 9Ls. Raytheon meanwhile struggled with quality problems as the AIM-9L nearly failed its second operational evaluation test.[14]

An "Interim Agile"

By January 1975 the AIM-9L was ready for joint navy–air force evaluation. That month five missiles were fired against QF-102 drones—planes that, ironically, had been designed to fire Falcons. A total of forty-six firings took place, the last on 1 March 1975. In 1976 Raytheon and Ford Aerospace put the missile into production, ending years of separate navy and air force procurement. The joint service missile was most critical for the air force, which finally got a missile with the range of the AIM-9D and the new all-aspect capability.[15]

The AIM-9L was an "extraordinarily lethal weapon." Tom Amlie called it "a death ray." The 9L's double-delta canards gave it an incredible 35-G capability, almost five times what average pilots can with-

Walt Freitag during rate-bias tests with Raytheon, late 1960s. Courtesy of Edward Paul

Mickie Benton during rate-bias tests with Raytheon, late 1960s. Courtesy of Edward Paul

stand, making it virtually impossible to out-turn the missile. Air force colonel Walter Vrablic told Raytheon's Russ Whynot that "when China Lake invented the 9L, I knew it was time to get out of the fighter pilot business." Combat performance of the 9L bore out this evaluation. During the 1982 Falklands-Malvinas War, well-trained Royal Navy pilots using AIM-9Ls chalked up an outstanding 87 percent success rate against their Argentine opponents. The Israelis racked up similar scores in the Bekaa Valley in 1982.[16]

AIM-9M

Dick Schmitt characterized the 9L development as "a somewhat hectic program." As with earlier models, not everything that could go into the missile went in. When the Defense System Acquisition Review Committee reviewed the program, it asked what additional improvements might be made. The air force, in particular, wanted better defense against countermeasures. These shortly became a product improvement package for the 9L, and the AIM-9L PIP turned into a new missile—the AIM-9M.[17]

In addition to counter-countermeasures, the 9M had better performance against cloud and terrain backgrounds. Optical filters had provided some resistance to countermeasures, but flares with the right optical signatures could defeat them. The new microchip electronics gave the missile enough processing power to sort out false targets, both background and countermeasures, from the real ones.

Funding remained a problem, but creative accounting provided enough to keep things going. Norm Woodall states bluntly, "There would have been no 9M without bootlegging." Trust and close cooperation between Ed Paul at Raytheon and Woodall at China Lake meant that decisions were made quickly and producibility was designed in. The AIM-9M went into production in 1981, and modified versions are in use today.[18]

Even though China Lake continued to support the missile, the brilliant 9M design team drifted apart after the project was completed. With the Family of Missiles Agreement, and ASRAAM (advanced short-range air-to-air missile) development assigned to Europe, designers may have felt there was little room for creativity. Whatever the cause, the breakup of the team ended the continuity of the Sidewinder

AIM-9L *(left)* and AIM-9R. Courtesy of China Lake Naval Weapons
Center

design effort. Some have speculated that the breakup was a factor in the
later demise of the 9R.

AIM-9R

International politics set the background for the 9R development. The
United States had signed the Family of Missiles Agreement with a
British-German consortium stipulating that the next NATO short-
range missile, ASRAAM, would be developed by the Europeans, while
the United States would develop the new medium-range missile,
AMRAAM. Hughes Aircraft eventually got the AMRAAM development contract and developed a good, but expensive, missile—the
AIM-120. For China Lake, however, the agreement had powerful
implications. The 9M Sidewinder team, which had absorbed many of
the Agile project engineers, was dissolved. The loss of this pool of brilliant young engineers would tell. China Lake had to defer tinkering
with the Agile technologies while it navigated somewhere between
AMRAAM and ASRAAM. China Lake would build a team virtually
from scratch to create the AIM-9R.

The AIM-9R would be China Lake's last major Sidewinder effort. Up to this time Sidewinder's elegant mechanical design had allowed it to operate with a free gyro and relatively modest computing power. In the new digital era, however, some designers thought the missile could dispense with its free gyroscope and use instead a sophisticated imaging system with vast computing power. The key to the new system was a charge-coupled device (CCD) detector. Instead of focusing on a point of infrared radiation, the CCD would allow a missile to "image" the target. Good optics and raw computing power would identify targets at long distance and sort out decoys.

Originally the CCD "camera" would have meant using the CCD as a simple detector. George Teate and others quickly worked up this idea. Done in true skunk works style, it was an attempt to *evolve* the 9M Sidewinder. China Lake devised two sets of flyby tests, Pave Prism and Long Jump, to select a CCD contractor. Pave Prism attracted Hughes, Ford, and Raytheon, but basically nothing passed the tests. The Long Jump Tests, starting in 1984, were undertaken at a White Mountain laboratory owned by the State of California. They were an attempt to get above the air pollutants at 12,500 feet, and they involved a great variety of targets—fighters, bombers, helicopters, even an air force airborne warning and control system (AWACS) E-3. The contractors, who were charged for data collection on only one of the tests, were delighted to get the data from all the tests.

Many feel that if China Lake had simply ordered production of the CCD camera seeker head on a 9M frame, the 9R would have been a success. But this was not to be. A group led by engineer Ken Banks decided the 9R would be radically different—in effect, a new missile. The attraction of the new CCD technology was strong, and China Lake agreed to incorporate it into a working missile along with more sophisticated signal processing, less drag, and longer range. A missile whose head-on capability could compete with Sparrow III seemed within reach. China Lake decided to push several frontiers at once— a big risk. In retrospect, it is clear that getting the most out of CCD technology and getting a working missile were contradictory goals. The China Lakers, furthermore, set themselves up when they told Congress that a production CCD missile could be developed for only 25 percent more cost per unit.[19]

The wheels of the gods now began to grind. In 1983, after eight years of experiments with CCD, the navy had decided to prototype an AIM-9M PIP. Three years later, China Lake persuaded the navy to accept full-scale development, transforming the AIM-9M PIP into the AIM-9R; on 19 April 1986, the navy issued a request for proposals (RFP). Secretary of the Navy John Lehman and Assistant Secretary Melvyn Paisley specified a fixed-price contract for full-scale engineering development. This might have been a sensible approach for developing the AIM-9M PIP—an evolutionary step—but the 9R was revolutionary and there were many uncertainties. China Lake wanted the project funded as a "feasibility study/engineering development," but Lehman insisted on a fixed-price plus incentive contract because the project was so far along. China Lake felt trapped, and the unworkable contract was born. The AIM-9R never should have been treated as if it were ready for engineering development; it should have been a feasibility study.[20]

In August 1986 Raytheon and Ford Aerospace submitted bids. Ford "bought in" and bid low, about $37 million. Raytheon took a different approach. After careful consideration of the RFP, Raytheon's response included many suggested changes—top management felt that the navy was asking too much for the price. Its bid included dramatically fewer features and asked for more money: $54 million. Ed Paul, the Raytheon Sidewinder manager, felt Raytheon could do the job, but not with the fixed-price and open-ended contract provisions the navy wanted.[21]

Raytheon's attempt to change the contract terms did not work. Capt. Jess Stewart, the navy program manager, said he could not have signed a contract containing such clauses. Still, despite the higher bid, he felt Raytheon had a stronger offer. The real cost of development, one observer has suggested, would be more like $150 million.[22]

The navy nonetheless accepted Ford's low bid.[23] Ed Paul felt this decision was a personal defeat. He thought it was also a fatal mistake for the navy, because Raytheon had tremendous experience to contribute to the new missile. Raytheon's top management, however, refused to lower the price or to sign off on contract clauses that threatened "open-ended losses." To Captain Stewart it rapidly became obvious the navy had made a serious mistake.

Ford used a bare-bones staff and had trouble lining up vendors who could handle the exotic parts. New management took over Ford's missile operations and was appalled at what the firm had agreed to do. Interest in the contract dropped. Stewart felt that Ford had a take-it-or-leave-it attitude, which left the navy little leverage.

China Lake had its problems, too. As opposed to the 9L project, whose integrated team had worked so well, the 9R mechanical, electronic, and software design teams had separate quarters. The new missile would stand or fall on its software, and Michelle Bailey, in charge of software, worked closely with the electronics section. But mechanical design somehow got left out of the team. A strong personality, Bailey sometimes failed to share information, for instance, leaving mechanical design uninformed about the test schedule. There also were conflicts about what kind of missile should be developed. While the mechanical team was arguing for producibility, Bailey and project manager Ken Banks set their sights on performance—and for a reason.

Banks had promised stellar performance for the AIM-9R. Yet while technically gifted, Banks had little experience at running a design project, and his promises proved impossible to keep.[24] NavAir expected a workable design at low cost per unit. This was unlikely since the 9R was a cutting-edge system. While the CCD was a wonderful image-generator, getting software to make sense of these images proved difficult. To some observers it seemed that "patches on top of patches were used to get configuration control," but Michelle Bailey denies any such lack of discipline on her team.[25] Both the navy and Ford, however, thought Banks's team was too highly focused on the testing program and neglected its role as a forerunner of production. Apparently, the developers "forgot that the end product was not a piece of hardware, but a document package." Deadlines kept slipping.[26]

Costs, meanwhile, began soaring. Ed Paul observed that one of the designated parts—the dome—had increased in cost by a factor of fifty; a two-axis gyro unit for the focal plane array cost $10,000. Projected unit costs—$70,000 to $180,000—began to approach those of the Sparrow. The air force was not part of the AIM-9R project, but it did not like what it saw. The missile was not a "front-line weapon" because its visible-light seeker could not be used at night. And it was complex. Engineer David Kurdeka estimated that there was "an order of magnitude difference" in the parts count between the 9L and the 9R. The exotic optical system was expensive to develop and would have been difficult to mass-produce.[27]

An equally serious problem was relations with the project pilots. While the 9L pilots had worked hard to make that system a success, the 9R pilots did not get along with Michelle Bailey. Project management had communicated that it wanted no test failures. The pilots interpreted this as pressure for perfect test results. Thus a constructive dialogue about the missile's problems did not materialize between designers and testers. Instead of socializing with the project team, the pilots tended to socialize with the 9M pilots. Furthermore, pilots did not seem to respect the digital missile and made few efforts to sell it to Washington. Since test pilots' opinions are influential, these defections hurt.[28]

Designer Lou Covert was the associate technical manager for the 9R. Covert, who had worked on the 9L, was frustrated by the emphasis on successful tests versus getting a finished package. The successful prototypes were not reflected in the drawing package sent to Ford from China Lake. In fact, two major fixes were required to get it working. Ford Aerospace began to grasp that it was facing financial disaster. Relations with China Lake soured.[29]

By the end of the project, Bailey's software was working. The first sixty-five seekers were delivered to the navy in May 1990, and AIM-9R development rounds were hitting their targets with regularity; five of the first six flights were successful. China Lake expected the AIM-9R to enter pilot production in late 1992, but success came too late. First the air force, then the navy grew restless with the many delays and slipped deadlines, and the navy canceled the AIM-9R in December 1991. Instead, of AIM-9R, "mods" of AIM-9M would continue in use through the 1990s.[30]

With the luxury of hindsight, it is obvious what went wrong. The AIM-9R technology should have remained a feasibility study for many more years before full development. A different contract between the navy and the developer should have been set: the attempt to develop a production missile was premature. Many people thought the development contract should have gone to Raytheon. Alternatively, a less ambitious use of the charge-coupled device might have allowed a simpler, less expensive round to be sent to production. The political ramifications of this failure would be grave.

The AIM-9M(R)

In 1988 the AIM-9M was getting long in the tooth. ASRAAM was not ready and the AIM-9R was struggling. At the same time, the air force decided to see how its much-tested inventory of Sidewinder AIM-9Ms would do against acquired Soviet decoy flares; they did not do very well. "Either by design or poor quality, these flares 'lit up' very slowly, and spoofed the counter-countermeasure design of the 9M."[31] [203]

Bill Boatright was then head of China Lake's air-to-air missiles division. His colleague Bob Pike stated bluntly that if the problem with slow-burning flares could not be fixed, "Sidewinder would go away." But Pike went on to point out that if the problem could be identified by simulation, then there was some way to fix it.

Boatright was intrigued and immediately summoned the key designers to a briefing in the Lauritsen Lab, where he told them they had to develop a solution or "the program goes out the door." There was silence, but eventually Schmitt and George Teate spoke up. The automatic gain control (AGC) added to the AIM-9L (and thus the M) might be the cause. Schmitt thought that if some of the AGC features were removed, the flare problem could get fixed in time for Desert Storm. Schmitt and George Banura worked on a fix that went into the AIM-9M-6/7.

Meanwhile, the Pentagon had convened a contractor meeting at China Lake to address the problem. All firms agreed that the right solution was not to fix the AIM-9M but rather a "total replacement of the front end [seeker]." Many felt that a new guidance-and-control section would overcome the AIM-9M seeker's attraction to slow-burning flares. Of course, each firm wanted to design and build the new front end. Air force planners, who still wanted a complete replacement, favored a rosette scanner similar to that used in the Stinger ground-to-air missile. From 1988 to mid-1990, the air force fought China Lake on the issue. Then in the summer of 1990, four days after Saddam Hussein invaded Kuwait, the air force suddenly became anxious to get China Lake's AIM-9M(R) prototype into production.

Between August and December 1990, several test firings took place—mostly direct hits—followed by operational evaluation with both navy and air force participating. The missile was turned over to Hill Air Force Base for serial modification, and sixty rounds of the AIM-9M-6/7 were produced; logistics problems precluded their use in Operation Desert Storm.[32]

Once the crisis was over, Schmitt started a small, co-located team under Norm Woodall's direction, using funds from the NavAir PMA-259 program. Using a Silicon Graphics computer, they simulated the missile's signal processing with different reticles. They took the 9M's FM scanning and paired it with the 9B's AM scanning and installed a logic circuit that told the seeker which mode to use. Two firings showed that the system would work. This became the AIM-9M-8/9.[33]

The entire AIM-9M(R) development process, including OpEval, had taken five months. Beginning with Boatright's meeting at Lauritsen, it had taken a total of twenty-four months and $26 million to go to full-scale production. This was a very cheap fix, but it was a fix for an old system. Almost five years earlier, in 1986, the Soviets had fielded an off-boresight, helmet-aimed, and thrust-vectored missile: the R-73 Vympel (NATO designation: AA-11 Archer). In short, the Russians had developed something like the long-sought Agile.[34]

The success apparently was caused by a mistaken intelligence estimate. When the United States canceled Agile and continued to use aerodynamic (canard) guidance, the Soviets assumed this was a cover story for making Agile into a "black" program. So they hustled to develop something equal to it. In reality, they pulled ahead of the United States. Soon, others would develop their own next-generation missiles.[35]

AIM-9X

The AIM-9R failure had severe consequences. China Lake seemed to have lost the magic touch. Many firms, such as Raytheon, Hughes, and others, had fine design teams; there was no reason, these firms argued, for China Lake to continue in the development business. The new challenge seemed to be the "off-boresight" missile. Using thrust-vector control (TVC) the next generation of air-to-air missiles would be capable of almost square turns and would be guided by helmet-mounted sights.[36]

During the 1980s the United States had left major short-range air-to-air missile developments to its NATO partners. By about 1990 it became obvious the Europeans were far from fielding their promised short-range ASRAAM.[37] (It may yet reach the field.) Only at this point did the United States decide to begin developing another short-range missile. Designated the AIM-9X, it is to be (once again) the last Sidewinder.

Hughes Aircraft beat Raytheon for the navy's $169 million contract to develop the AIM-9X. Shortly after that, Raytheon bought Hughes. China Lake was a contender for the airframe only and offered a concept called Boa; the Department of Defense decided not to use it. In principle, the AIM-9X will be the perfect dogfight missile. It is expected to go into production in 2002, seventeen years after the former Soviet Union's R-73 Vympel. China Lake's role will be minor.

Foreign Copies, Improvements, and Alternatives

The American success in developing air-to-air missiles changed air combat. Sidewinder was the great success, and other nations faced the choice of buying or copying it—or designing a better missile. While Sidewinder did not have the longest range or the biggest warhead, its simple design made it hugely popular, not only with the United States and its allies but also with its opponents. U.S. allies bought Sidewinders, but many wanted to improve it. France, Germany, and the United Kingdom soon developed alternatives, many equal to the American concepts; many of the British concepts got squashed by internal politics.

Others took Sidewinder's basic concept and improved on it. For instance, Israel's Shafrir (later Python)—introduced shortly before the AIM-9D—was an excellent short-range missile. Looking very much like a fatter Sidewinder, with similar configuration and guidance system, the Shafrir is credited with destroying more than two hundred enemy aircraft. Similarly, the French firm SA Matra developed the Magic, a missile whose specifications were even more demanding than those of Sidewinder.[38]

The Russians Get Sidewinder

Sidewinder's design could hardly stay secret long. Widely distributed to U.S. and other NATO forces, a Sidewinder was soon in the hands of the former Soviet Union. The Soviets copied Sidewinder first, then redesigned and improved it. Some have claimed that the first Sidewinder to fall into Soviet hands arrived during the second Formosan

Crisis, when a Sidewinder 1 that hit a Chinese Communist MiG failed to detonate. Stuck in the MiG's wing root, it was carried back to the mainland, where the plane landed safely; soon, it was in the Soviet Union.[39]

What is certain is that the Soviets obtained the plans to the AIM-9B (Sidewinder 1A) early in the game from Col. Stig Wennerstrom, a Swede. The plans even included Jim Madden's washing-machine checkout system.[40]

Soviet reactions to the missile were powerful testimony to its value:

> During the first several months of study and analysis of the missile, the Soviet participants were in a simultaneous state of euphoria and technological humbleness. *The almost biological harmony of all the functional elements of the missile, the elegance and the powerful logic of all the aspects of the Sidewinder engineering were obvious and overwhelming.* The name of Mr. McLean, the American Chief Designer, was well known to the OKB engineering staff, as was the fact that he had received a presidential award for the development of the Sidewinder. By comparison the K-5 was considered by most members of the task force to be hopelessly obsolete, although this attitude was subsequently modified and the K-5 R&D resumed.[41]

In spite of the Sidewinder's advanced design, the Soviets were successful in replicating it. The first missile—given the code name Atoll by NATO—was built to Philco specifications so closely that it used the Philco parts numbers. Since that time, the Atoll has been continuously improved. NATO observers first saw the Atoll on a Soviet aircraft on 9 June 1961 during an air show. During the Vietnam War, Atolls would hit American planes with lethal effect.[42]

[16]

In Combat

The new guided missiles changed air combat forever. With both range and self-guidance, missiles were now the dominant weapons system. If properly used, they could even down an enemy beyond visual range. Glenn Tierney was only slightly exaggerating when he said that "a pilot with a Sidewinder is like a fighter going into the ring with a six-foot reach." Yet the advantage missiles conferred was relative, not absolute. No missile hit the target every time. In Vietnam, for instance, while some missiles did down enemy aircraft, others failed due to mechanical or tactical problems.[1]

Combat is the toughest test a weapon system gets. In combat the theories of the weapon-builder are matched against actual results. Combat proved a sobering experience for the air-to-air missile designers. Sidewinders had been shown to work in hundreds of tough trials, including those by the fleet. But combat was still more demanding. Maintaining missiles on carriers or in tropical bases caused big reliability problems. Then there was hitting the target. In combat the target maneuvered—and shot back.

Perfect or not, no pilot wanted to launch without missiles. Glenn Tierney's "six-foot reach" was an advantage no one wanted to yield. No one thought of going back to guns and rockets.

Pretest over the Formosa Strait

Before the U.S. fleet got a chance to try out its new missiles in combat, the Nationalist Chinese entered the lists. In the second Formosa crisis the United States aided the Nationalist forces of General Chiang Kai-Shek in the struggle over China's offshore islands of Quemoy and Matsu. The U.S. Navy assembled an awesome naval force in the area. This force included the carrier *Midway,* which was armed with nuclear weapons. Rather than use its military might directly, however, the United States decided to bolster Chiang Kai-Shek's forces with some of the latest hardware—including Sidewinder missiles.[2]

A small five-man team from China Lake, led by Cdr. Sel May, put together Operation Black Magic. Some 360 Sidewinders were used to arm one hundred Chinese Nationalist Air Force (CNAF) F-86s, giving them an ability to pitch up and shoot at the higher-flying MiGs flown by the Chinese communist air force.[3]

The team quickly brought the CNAF F-86s into the missile age. Nor did they have to wait long for results. On 22 September 1958, a force of ten CNAF F-86s took on perhaps double that number of MiGs and routed them. Among the triumphs was the shooting down of four MiGs using Sidewinders. The Nationalists held a big banquet to fete their victory as well as the advisers from China Lake. Sidewinder worked in combat.[4]

Air War Vietnam

With such a good beginning one might expect that "the missiles worked happily ever after." This was far from true. In Vietnam the United States encountered severe problems using air-to-air missiles. These included problems with pilot training, tactics, maintenance, manufacture, and design. As the war progressed, some of these problems got fixed. Others, however, proved difficult to correct. What were these problems?

1. The most important problem was that the missiles failed to work properly. By the end of the war, one expert calculated, of the 1,066 Sidewinders and Sparrows that were fired, 617—a whopping 58 percent—failed to function. Sidewinder (47 percent failures) was better than Sparrow III (66 percent failures), but certainly was nothing to brag about. These high failure rates represented both poor manufacturing and poor maintenance.[5]

2. Many missiles were fired out of envelope. Many pilots did not know what the proper firing envelope was, and for the Sidewinders, the pilots

often could not easily determine from their cockpits if the target was in the envelope. And some missiles were fired as part of a salvo or to scare the enemy.

3. The missiles had been designed as antibomber weapons but were used against fighters. Sparrow, Falcon, and Sidewinder had all been designed for the antibomber role. But their adversaries in combat were typically highly maneuverable fighters, MiG-15s, MiG-17s, and the very fast MiG-21. A sharp turn "pulling G" could often shake off the missile. (This was also true of the Atoll missiles fired at U.S. aircraft.)

[209]

4. Finally missiles used in combat are part of a vast system using pilots, aircraft, radar, tactical intelligence, and organizational learning. Without tuning up the whole system, many advantages that the missiles might have provided were lost. The needed cooperation came along only slowly and became good only toward the end of the war.

Why did these problems occur?

Underlying Causes

The roots of the missile problems went deep into the history of the missiles themselves. Sidewinder, for instance, had been designed by a community that believed the missile made aerial guns and rockets obsolete. This community had tested the new weapon under a variety of conditions, some stringent, but none as stringent as combat. It had sold both Washington and the fleet on the idea that missiles and only missiles were needed for aerial combat and that they would work reliably. This was what the community itself believed. Bill McLean refused to accept the idea, for instance, that Sidewinders would not hit the target nine times out of ten, yet this confidence was based on theory not combat experience.

Of the three missiles used in Vietnam by the United States, Sidewinder was clearly superior. Falcon was introduced late to combat forces and was quickly proven inadequate. Sparrow III was deadly when it worked but often had crippling quality problems. Sidewinder, whose development had cost one-tenth that of Falcon or Sparrow III, was more reliable.[6]

Yet even Sidewinder required skill. If the properly trained pilot used a properly made missile within the right envelope, more times than not the target died. Unhappily these three conditions were seldom united. Either training or manufacture or judgment failed. The missile missed, and the foe lived on to try his missiles or his guns on the shooter. Of every five Sidewinders fired, only one hit.

Sparrow III was worse. Raytheon had supplied the mainframe computer for Point Mugu, had test instruments that were superior to those of the navy, and controlled many of the conditions of testing and evaluation.[7] The "gold-plated" and carefully crafted test rounds had far greater reliability than the production rounds had and the "middle of the envelope" tests were nothing like as tough as those Sparrow faced in combat. Falcon's testers were equally optimistic, with just as little reason for optimism.[8]

So much for the United States' missiles. The fighter aircraft community that would use these missiles was also, on the whole, rather unprepared for combat that included air combat maneuvering (ACM). Only the navy's F8 Crusader, a pure air-superiority machine, was designed for ACM combat, with Sidewinders and a pilot community that was well-trained in fighter maneuvering. The Crusader, an aging but effective machine, also had a Vulcan cannon, which would be put to good use, even in a missile-firing environment. Many navy Crusaders would fire their AIM-9B and AIM-9D Sidewinders effectively against the MiGs or use a cannon to polish off the fight if the missiles failed.[9]

The air force had the new F-105 Thunderchief, a powerful fighter-bomber, able to carry a hefty bomb load and sporting an excellent cannon but unfortunately a poor gunsight. While the "Thuds" also had high speed on their side, the F-105 was hard to maneuver. At first it was not equipped with missiles. Since the Thuds handled the majority of early bombing missions, they also got plenty of experience with MiGs. Thuds posted twenty-seven MiG kills.[10]

The other new aircraft was the F-4 Phantom II, a plane designed for fleet defense and equipped initially only with the Sparrow III. Sidewinders were added later and eventually guns as well—a feature designers didn't feel was needed on the Phantom and which some had tried to remove from the Thunderchief as well. Unlike the Crusader community, the Phantom drivers had been trained to believe that their Sparrows would knock down foes beyond visual range, making dogfights unnecessary. They had therefore received little training in air combat maneuvering, and in fact little was known of which ACM tactics worked for the Phantom II.[11]

As the first part of the Vietnam air war (Rolling Thunder) unfolded, these were the kind of air-to-air assets the United States had. The U.S. weapons community had also made a key assumption: There would be no more dogfights. When you acquired the target, you locked on your missile, pressed the button, and the missile did the rest.

There were two things wrong with this assumption. Problem number one was believing that the pilot would know what he was shooting at. IFF (identification friend or foe)—electronic identification of other aircraft—was expected to work, but for a variety of reasons it did not. After the first "friendly" was shot down by a Sparrow III, the rules of engagement were changed to permit firing only on targets that had been visually identified. This nullified much of Sparrow's beyond-visual-range advantage.

Furthermore, dogfighting favored the closer-range Sidewinder over Sparrow. Sidewinder had the advantage in a dogfight because it was easier to ready, simpler to understand, and more reliable once it was fired. But (problem number two) Sidewinder could not work in the head-on mode but only in a tail-shot. This meant the shooter had to get behind the target. Getting behind meant maneuvering, since targets seldom obligingly flew in front of the shooter's aircraft.

Thanks to the "no-dogfight" assumption, and the design history of the missiles, pilots often flew into battle with weapons and training designed for another mission.

Yet that is pretty much how it is in a war. The equipment is seldom exactly right. The training is usually incomplete. The tactics are often unrealistic. As with missile development itself, combat success comes not so much from what you start with but rather how quickly you can improve it.

Combat Experience

Combat experience soon showed the complexity of fighting with guided weapons. Far from the push-button war envisioned, the pilot found himself in a dynamic environment where many traditional skills had to be adapted to the new weapons. Even for the experienced, missile combat required adaptation. For instance, early in the war, on 10 July 1965, four air force F-4Cs from Ubon Royal Thai Air Base, Thailand, decoyed a group of MiG-17s from Phuc Yen airfield into an attack. Lack of certain identification of the enemy meant the Sparrows could not be used beyond visual range. When the MiGs got within Sidewinder range, the flight of four Phantoms split up, thus two separate engagements took place.

> The MiGs' noses sparkled with bursts from their lethal 23mm and 37mm cannon. Holcombe and Roberts [pilots of two separate F-4s] accelerated and turned in opposite directions, splitting the MiG attack further. A series of wild gyrations followed, with Holcombe using both his high-g turn and vertical capability first to shake off the MiG attacking him and then launch a head-on attack. This could have been fatal, for Holcombe and Clark [Holcombe's back-seater] suffered some confusion over the armament setting and had to endure another firing pass from the MiG while they got their signals straight. Then, following the now classic energy-maneuverability formula [developed by Col. John Boyd, U.S. Air Force, assisted by civilian Thomas Christie], Holcombe, in full afterburner, went into a 10,000 foot dive and then high-g barrel-rolled through another MiG gun attack to wind up behind the quarry and within Side-winder range. *Four Sidewinders were fired: three seemed to miss* and the fourth had not impacted when the MiG entered a cloud layer. Holcombe and Clarke were sick with disappointment until they heard that the No. 2 F-4 Phantom had observed their target blow up.
>
> Roberts and Anderson in the No. 4 aircraft did not have quite the same tussle with their MiG-17. They made a textbook dive-and-climb attack which positioned them behind the enemy, and got off three Sidewinders. *The first two missed* but the third one blew up just behind the MiG, which went vertically into the ground. It was a good beginning to what was to be a long war.[12]

As the quotation shows, dogfights, salvos, and misses were common. And this was a lucky engagement, since in this case both MiGs were dispatched. Things did not always go so well. Often all the missiles would be fired without effect on the enemy. Or one of the enemy's missile shots or gun bursts caught the U.S. fighter. The situation could get grim when long periods went by without any successes. In June and July 1968 some thirty Sparrows were fired by aircraft launched from the USS *America,* with only one hit.[13]

Combat experience thus showed that the "no dogfight" assumption was wrong. Two early successes with Sparrow III had seemed to suggest otherwise, but they were soon followed by a shoot-down of a U.S. aircraft with another Sparrow III. Clearly, firing missiles beyond visual range, in the complex conditions of combat, was dangerous. IFF was not going to work. In the early part of the war, at least, one had to get close. And getting close meant the dogfight, supposedly banished forever, was still a going concern. Pilots learned to survive in dogfights or they didn't survive at all.

[213]

Combat experience was different for the air force and the navy, and it was also different for the three different U.S. fighters. The navy did better in combat because it had better weapons. Naval aviators often flew the F8 Crusader, a plane designed for the air superiority role, whereas the air force flew the Phantom II and the Thud, both designed as fighter-bombers. Phantoms, at first, also had no guns, which put them at a disadvantage against Russian fighters that did. Both Crusaders and Thuds had guns, and used them with lethal effect. When guns were installed on Phantoms, their pilots also made gun kills.

Nor did it take long for the pilots to recognize that Sidewinders were usually a better option than Sparrow III. In addition to higher probability of kill once launched, Sidewinders were just plain easier to use. When Falcons became an option for the air force, experience quickly showed the difficulties in using the compact but complicated weapon. Parallel experience with the ineffective E and J versions of Sidewinder actually steered air force pilots back to use of Sparrow III during Linebacker (the second phase of the air war). Many air force pilots wanted the navy AIM-9D version of Sidewinder, but launchers for AIM-9E and 9J would not take the AIM-9D.

When Sidewinder worked it was deadly. The shooting pilot waited to hear the "growl" in his earphones, indicating that the seeker on the 'Winder had acquired the target. Then he pushed the pickle button and said "Fox away!" or "Fox Two!" as he fired. After about half a second, the missile left the launcher, dropped slightly, then pursued the target. In about two or three seconds, the missile arrived. Sometimes the Sidewinder would go up the target's tail, blowing the aircraft into small pieces. More commonly, the missile would pass close by, detonating the proximity fuze. The exploding warhead would then cut the target apart. Lt. Philip Wood describes an engagement in May 1967. After being attacked by a MiG-17, Lieutenant Wood pulled around the MiG in a fast turn.

> [The MiG driver] must have been an inexperienced pilot. He bugged out by reversing course, which allowed me to park behind him at 2,000 feet. He was running for his life. I switched my stick's armament switch to "heat" and put him in my gunsight. The missile growl was loud. I pulled the trigger and the half-second before my second Sidewinder fired seemed like the proverbial eternity. As the 'Winder left the rail it dropped and appeared to have no guidance on the target. But moments later it started a gentle glide toward the bandit's tailpipe. The missile impact and the expanding rod warhead cut his entire tail off. The MiG lazily pitched nose-over and decelerated rapidly.[14]

Again, this was a happy ending (for the U.S. pilot). Something of the overall success rate for Sidewinder and Sparrow can be grasped from tables 16.1 and 16.2, representing two phases of the air war.

Sidewinder thus provided a more certain hit. These figures represent improvement over time, which reflects both better training and improvements in the hardware itself. Later versions of Sparrow and navy Sidewinders had higher kill ratios. During Linebacker the worst Sidewinder was the air force AIM-9E, which had about a 9 percent SSKP (the improved 9J was better with a 13 percent kill rate), and the best was the navy AIM-9G whose rate was a whopping 46 percent.[15]

The Americans' advantage was not only better weapons but also better training. The North Vietnamese advantage was that most American flights were on strike missions over territory watched over by North Vietnamese ground controllers. Thus initiative for attacking was almost always on the North Vietnamese side, with the American fighter escorts having to respond. Furthermore, thanks to Russian and Chinese hardware and advisers, the North Vietnamese had a highly integrated and sophisticated ground control system.

Table 16.1. Rolling Thunder (1965–68) Missile Results

	Fired	Failures	Misses	Hits	Kill Ratio[a]
AIM-7 Sparrow III (D & E)	340	214	99	27	8%
AIM-9 Sidewinder (B & D)	187	105	53	29	16%

Source: Adapted from Marshal Michel, *Clashes: Air Combat over North Vietnam, 1965–1972* (Annapolis: Naval Institute Press, 1998), 151–54.

[a] Hits/Shots

Table 16.2. Linebacker (1971–73) Missile Results

	Fired	Failures	Misses	Hits	Kill Ratio[a]
AIM-7 Sparrow III (E)	272	190	53	29	11%
AIM-9 Sidewinder (D & G)	267	108	107	52	19%

Source: Adapted from Marshal Michel, *Clashes: Air Combat over North Vietnam, 1965–1972* (Annapolis: Naval Institute Press, 1998), 151–54, 287.

[a] Hits/Shots

Air superiority, of course, was not an end in itself. The purpose of most U.S. air missions was *not* killing MiGs but air strikes. Air superiority was important because an effective MiG deterrent allowed strikes to proceed. A strike force was effective only to the extent that it could be protected from MiGs. It is true that most U.S. aircraft losses, about 90 percent, were due to antiaircraft ground fire and surface-to-air missiles (SAM). But this represents a condition shaped by effective defenses against MiGs. If U.S. forces had been less effective in dealing with the MiGs, there would have been more air-to-air losses from them.

Learning the New Air Combat

The United States was fortunate at the beginning of the war to have many experienced pilots, including such leaders as air force colonel Robin Olds, a World War II ace (he just missed becoming an ace in Vietnam as well). As the war continued, however, through attrition or rotation, these experienced men were replaced by pilots with less combat time and thus less likely to survive in combat. But the biggest problem was simply that the United States had not carefully studied and taught the tactics for the new air combat using missiles. Many pilots fired their first Sidewinder in combat. They often had little idea of its limits and capabilities. This was also true of Sparrow III.

Every weapon requires skill to use. Skill becomes possible only with experience, but the experience needs to be written down, studied, codified, and carefully taught. This seldom happened, and it never happened systematically until the navy created Top Gun. How Top Gun took place is an interesting story in itself, which I can tell only in part.[16]

A carrier officer, Capt. Frank Ault, became concerned about the relatively low combat exchange ratio between the United States and North Vietnam. He wrote many letters to Washington but without result. Not until a rotation to Washington in 1968 was Ault suddenly summoned to a meeting with Rear Adm. Bob Townsend and Vice Adm. Tom Connolly. These men were, respectively, commander of Naval Air and deputy chief of Naval Operations for Air. Ault found himself explaining why the air-to-air problems needed fixing.

> To my surprise I found that I had talked myself right into the hole that Townsend and Connolly already had dug and readied for me. I was given the "simple" mandate to find out why the navy was not shooting down more MiGs and to recommend what should be done about it. I was given *carte blanche* to pick anybody I wanted to assist and to conduct the study in any manner I saw fit. As for desired end results, I was told "an increase in combat kills of not less than three times better than performance to date, and as soon as possible."[17]

With this urgent assignment in hand, Ault handpicked his team and got them working. The team, led by experts in five areas (design/manufacture, maintenance, fleet support, training, and missile repair) flew many miles and got powerful industry support. In three months they compiled a report recommending changes in every area—242 of them.[18] One of these recommendations was the creation of a "graduate school" to teach fighter weapons tactics. Through a process that would have made China Lake proud, and against some resistance, a fighter weapons school, soon to be known as Top Gun, was created near San Diego. Graduates of this school fought in the second phase of the war. They did extremely well.[19]

As Marshall Michel shows in his book *Clashes,* there was also constant learning about tactics on both sides, as move and countermove succeeded one another. The missiles themselves improved, both as a result of the Ault study and also as a result of constant feedback from the air war to manufacturers and to the service desks. The navy went through three Sidewinder models, so did the air force (the navy Sidewinders improved faster). Sparrow III went through two models and many "mods." Falcon was tried and found wanting. Designers responded to pilots' pleas for guns on the Phantom II. And increasing use of airborne radar warnings systems led to improvement of intelligence in the skies. The United States got two missile aces (one with Sidewinder, one with Sparrow III).[20]

The air force did not learn its lessons about weapons, training, or tactics as fast as the navy. Linebacker air force results with the AIM-9E were *worse* than with the AIM-9Bs used during Rolling Thunder. This may have reflected less experienced pilots, but it also reflected lack of attention to ACM skills. The air force did not establish its own combat maneuvering range at Yuma until the war was over. Similarly, the air force hung onto its "fluid four" formation long after the navy had given it up in favor of the "loose deuce." The reasons for these differences lay in the internal politics of the two services.[21]

Theory and Practice

The surprises the United States experienced in the Vietnam War revealed a basic fact of life: the difference between the laboratory and the battlefield. The laboratory is a controlled, linear environment, where inputs can be carefully monitored and shaped, and outcomes carefully calibrated. By contrast the battlefield is an uncontrolled, chaotic environment, where conditions change on a day-to-day basis. The Sidewinder team had used its full powers to orchestrate conditions that would allow the new missile to emerge from the laboratory and thrive. In Vietnam similar teamwork was needed to orchestrate conditions in the air to provide an optimal environment for using the new weapon and others like it.

As the Vietnam War progressed, learning took place. U.S. forces became more organized. Cooperation between combat operations and airborne intelligence gathering became closer. The equipment for the navy improved with better Sidewinders. The equipment for the air force improved with better Sparrows (air force Sidewinders were another story). The navy created Top Gun to train pilots in air combat maneuvering. Better coordination was achieved with airborne tactical intelligence. Many of these lessons, though, were learned at great cost in equipment and men.

The most important lesson of Vietnam was that missiles were good weapons but not superweapons. They worked if used correctly: when the right design was carefully manufactured, suitably maintained, and fired within a well-defined launch envelope. Sidewinder in particular was a good design, but it was capable of considerable improvement. Over the years, it got those improvements.

The Wars of Israel

Israel had its first missile kill on 28 November 1966, when an Israeli Mirage IIIC fired a Matra 530 missile successfully at an Egyptian MiG-19, downing it. Yet missiles were hardly used at all in the Six-Day War of 1967, whose many Israeli victories were all accomplished with guns. After 1967 there was a rapid climb in the number of missiles Israel used in combat. The Israelis soon got access to Sidewinders and used the AIM-9B and AIM-9D in the early years. By the Lebanon conflict (1982), Israeli planes were also carrying the AIM-9L "super-Sidewinder." Diplomacy got Israel its early missiles, but soon it was designing its own, first the Shafrir and then its successor the Python. The Rafael Design Bureau carried out the design of these Sidewinder variants. Israel has excellent technical capabilities, and these missiles are generally considered superior to the American Sidewinder versions.[22]

[218]

The Falklands Conflict

The Sidewinder AIM-9L played a key role in allowing England to dominate the Argentine air force in the conflict over the Falkland Islands. British air forces (Royal Navy and RAF) fired twenty-six AIM-9Ls during the conflict and were directly responsible for the destruction of eighteen Argentine planes. This is a 69 percent hit rate, which might even be set higher if one considers that in two instances of misses, a second round hit the target and destroyed it. The British had put their naval aviators through intensive training in the use of guided missiles. As a result, the air campaign was so successful that soon Argentine fighters simply would not engage British fighters. Unhappily this did not keep the Argentines from bombing British ships or using their small inventory of Exocet antiship missiles on them, with deadly effect.[23]

[219]

Gulf War

During the Gulf War Allied forces used the Sidewinder AIM-9M in the one-sided air battle. After a few early engagements, Saddam Hussein moved the Iraqi Air Force out of harm's way. Nonetheless, the allied forces managed to score twelve kills with Sidewinders and about double that number with Sparrow IIIs. The excellent IFF tactical intelligence available to the allied forces allowed use of beyond-visual-range missiles, particularly on fleeing Iraqi aircraft. While Sparrows performed better than expected, they were still not up to Sidewinder. And neither the Phoenix nor the AMRAAM, the new and expensive missiles, proved of much value. So the air battle was fought with updated versions of 1950s weapons.[24]

Conclusion

Battle is the fiery furnace in which theories about air-to-air combat are tested. Many of the assumptions taken into battle do not pass the test. For instance, the push-button war that many expected did not materialize in Vietnam. Pilot training was still critical. Even in the Gulf War, as close to the push-button ideal as one can get, ACM tactics still counted. Pilots had to be able to dogfight. Maybe they always will.

A missile, like any other system, is tested by experience. The creators of Sidewinder had tried to foresee everything. They did better than the other missile designers but still discovered, in the words of "Poor Richard," that "experience keeps a dear school."

[17]

Building People at China Lake

The creativity of an organization may come from down below,
but it has to be inspired and nurtured from above.

Mickie Benton

There was much more to China Lake than met the eye. Below the surface, hard to see but absolutely essential to the functioning of the center, were a series of informal structures that supported its operations. These existed on several levels and interwove with the formal structure in complex strands. Although this web of people and practices was less visible than organization charts and technological products, China Lake's creative culture depended on it.

Getting the Right People

With some five thousand employees, China Lake was a big laboratory. Getting the right people was critical, and China Lake invested heavily in recruiting. Because it could not match private industry's salaries, it had to be creative. While the old-boy network had provided virtually all the initial players, L. T. E. Thompson, the first technical director, made recruiting his first priority, and he set the tone for China Lake culture. Later practice maintained this standard. Les Garman, who served on the ten-member recruiting board and was its head from

1965 to 1973, emphasized that no pains were spared to recruit the best people:

Top managers strongly supported recruiting. Among them were such technical directors as L. T. E. Thompson, Fred Brown, Bill McLean, Tom Amlie, Walt LaBerge, Hack Wilson, and Leroy Riggs. Military commanders who supported the effort included Switzer, W. Vieweg, P. D. Stroop, D. Young, F. Ashworth, and Hardy.

[221]

The station's top talent often did the actual recruiting. Hack Wilson, as associate technical director, insisted on sending out the top bench scientists—"the ones who couldn't be spared for even a single day,"—because he knew they would provide convincing role models.

Travel funds were always available for recruiting, and China Lakers sought opportunities to spread the word by addressing students and faculty at places such as Caltech.

The personnel office generally backed up the judgment of any recruiter who felt that a young scientist or engineer should be hired on the spot—subject, of course, to a security check.

Recruiters received special training. Sometimes their pitches were filmed and their performances critiqued. China Lake recruiting became a model for other civil service operations; the Civil Service Commission in Los Angeles invited Garman in to give them pointers.

China Lake personnel were encouraged to contact their colleagues in academia to identify promising young students, who were then targeted for interviews; Ph.D.'s were invited to visit the station at government expense.[1]

Recruiters sometimes exaggerated the charms of desert life, but their promises of personal fulfillment and job satisfaction were often right on the money. So China Lakers invited former colleagues to join up. For instance, Bernie Jaeger arrived in 1948 after completing graduate study in aerodynamics at the University of Minnesota, where he had shared office space with Dale Drinkwater, who went on to Notre Dame. After trying to support a large family on an assistant professor's salary, Drinkwater called Jaeger; shortly after, China Lake hired him. Drinkwater's Notre Dame connections proved instrumental in hiring four more highly qualified Fighting Irish Ph.D.'s—including Walt

LaBerge. Drinkwater eventually became head of the supersonic naval ordnance research track (SNORT).[2]

Drinkwater also recruited Tom Amlie early in 1952 as Amlie was finishing his doctorate in electrical engineering at the University of Wisconsin and getting ready to go on active duty. Ensign Amlie had orders to submarine chaser school in San Diego, but Cdr. Thomas Moorer, then the experimental officer at China Lake, got his orders changed with a telephone call.[3]

Surprisingly, while every effort was made to get Ph.D.'s with top grades, the recruiters looked for engineers who were the consistent B rather than A students. China Lake viewed the A students as potential prima donnas, while the B students were seen as more likely to drive projects through to completion. Less surprising was the recruiters' interest in previous hands-on technology experience—such as working on cars or electronic tinkering.

Matching People and Projects

The system attempted to match newcomers with projects they liked. Thompson's Junior Professional Program (still in use) exemplifies the spirit. The newcomer selected programs and rotated to a new one every three months during the first year. Woody Woodworth entered the program just after getting his B.S. in physics, going first to Sidewinder and then to aerial fire control. He did not feel comfortable with Sidewinder but liked fire control so much he refused to continue with his program rotation:

> From my personal standpoint, the Sidewinder group was quite self-contained; they were very confident people. And my impression was at that time to continue in there, I would be very much of a small fish in a big pond. [So I went to aerial fire control.] The [fire control] group under Henry Swift was very unusual, and very competent. It scared me a little bit. I was this little kid from South Dakota State College of Mechanical Arts. There were all these people from Berkeley and Caltech, and Dr. Swift. . . . I enjoyed it, so I stayed there.[4]

Woodworth's branch head was Jack Crawford, a brilliant engineer for whom he developed immense respect. The two soon developed an intellectual partnership. Crawford had very strong physical intuition and "systems thinking" ability; Woodworth was a skilled problem solver. Their partnership eventually led to the Walleye glide bomb.

John Gregory, Woodworth's division head, understood his strengths and did not burden him with paperwork. When Woodworth was made a branch head—a seemingly natural progression—he immediately proved to have little talent for administration. So he went back to the bench, where he continued to be a star performer. China Lake's ability to handle people like Woodward goes a long way toward explaining its success.[5]

[223]

Work with Meaning

Sidewinder is a good example of how China Lake projects created meaning. The project made sense. Every hour put in after work, every postlaunch party, every interaction with McLean, Wilcox, LaBerge, or Ward told the members that they were part of something special. Working with McLean out on the desert long after dark, tinkering in the wee hours to make one last adjustment, knowing that the rest of the group depended on one's own work reminded the technician, engineer, scientist, or pilot that he belonged.[6]

McLean's goal was to get everyone involved with the project, whether civilian or military, government or contractor, to form "an intelligent set of people who would act as if badged by the lab," LaBerge said. McLean did not always convince the contractors, but otherwise he succeeded.[7]

There was a minimum of pomp and circumstance. The military observed normal formalities in transactions with each other, but the civilians largely ignored any hierarchy. World War II government laboratories had developed a tradition of omitting the "Doctor" when addressing Ph.D.'s; first names or nicknames were used extensively.[8]

McLean's appearance in the laboratory after hours and his constant presence on the firing ranges showed that no one was above getting his hands dirty. Equally important, however, was the constantly reinforced idea that the work was primary and that organizational structure and its symbols were secondary. Warren Legler notes that technicians were considered "other enthusiasts" in the hobby shop. After a test, a technician would toss a part on the test table and say, "Hi, so and so! Here's the umbilical cord from test number. . . ." While technicians did not usually socialize with the engineers and scientists at the Officers Club or the Elks Club, the overall effect of egalitarian practices encouraged everyone to identify with the work.[9]

The project also attracted people because it seemed likely to succeed. The psychology of teams that expect to succeed is different from the psychology of teams that expect to fail. The will to think, a phrase coined by Enrico Fermi, is created when the research will lead to concrete results. Fermi said that he was encouraged to proceed in the development of nuclear power because he knew the U.S. government would provide funding. This sense that his efforts would bear fruit gave him the confidence to see the project through. The same feeling permeated activities at China Lake.

The project began in wartime and continued during a period when war was a serious threat. Many of the team members were World War II veterans. What the station did often had a direct impact on combat. Feedback from the fleet that a weapon had worked caused rejoicing, and failures produced deep concern. Frank Knemeyer, a department head, commented:

> If you wanted to make a lot of money at your job, you didn't work for the government out here. So what you did get, you got the satisfaction you're talking about—that the thing worked the way it was supposed to. In Korea and Vietnam we actually helped pilots accomplish their mission and *saved their lives.* Carrier skippers and carrier air group commanders would come here before they deployed. We would spend a day here with them, telling them all about our systems and everything, everything we knew about their idiosyncrasies, and in some cases we helped train them here. We would also send people out to the carriers to help them whenever they ran into any problems, everywhere from the technical, even to helping them with their tactical problems. When those people came back from deployment they made a point to stop here and tell us how they made out, how our systems worked, what were the bad points, and basically, how it saved their lives. That is not a monetary reward—that is one of the richest rewards you can get.[10]

People who didn't understand the wide-open China Lake process often characterized its efforts as a hobby shop; they were off the mark, but the team did encourage personal tinkering as a release from day-to-day pressures. While McLean was busy creating the Sidewinder missile, he also perfected valves for scuba apparatus. Ed Swann, Jess Lamar, and Bob Meade competed to develop the ultimate lightweight pack frame. Lamar's quest for the ideal recurve bow was typical of after-hours pursuits. Ralph Dietz helped air force captain Tom McElmurry build an eight-inch reflector telescope in China Lake's excellent Optics Lab; Dietz had helped to calculate the curvature of the Palomar Observatory mirror. "Michelson Lab was an educational facility in the evenings," McElmurry observed.[11]

[225]

Fun was part of the process. Entertainment provided an important social glue. One of LaBerge's most important roles was to act as the group's poet and songwriter. The team produced four operettas that not only helped people relax but also provided an important chance to vent tensions and frustrations.

They partied hard. The 1950s-era AIM-9B Sidewinder team developed a tradition that continued into the 1970s. In the early days of Philco's involvement, the company's representative often threw barbecues with a wheelbarrow full of salad. Twenty years after the initial Sidewinder success, the AIM-9L "super-Sidewinder" group was still upholding the tradition. Lee Sutton, an electrical engineer just hired, went to a farewell luncheon at the Hideaway, a local watering hole, at about 11:30 in the morning. At 1 P.M., the group was still on its drinks and had not yet begun to eat. Many did not return to work.[12]

There was a party after every successful AIM-9L shot; in fact, there was a party after every unsuccessful shot. Parties included everybody, from engineers to missile handlers—the "white-hats." Glenn Hollar, the Sidewinder program manager, often paid for beer out of his own pocket. When Hollar left as program manager, his successor Allen Gates did not continue funding the parties, so Mitzi Fortune, and Shirley and Maggie Pladson started the Wuzwinder parties. These became so popular that tickets were used to limit attendance. People then resorted to forging tickets. One party sold 360 tickets but had at least 600 attendees. Washington sponsors and contractors scheduled visits to China Lake to coincide with the Wuzwinder parties.[13]

It is hardly surprising that so many of the team remember this as the best time of their lives; the phrases "the golden days," "the best years of my life," "magic," and "Camelot," recur in their accounts. It is impossible to mistake the feeling, so rare in the world of work.

The Technical Infrastructure

Just as there was an informal system for the promotion of ideas, there was an informal system for their implementation. Concepts began as sketches that had to be translated into drawings, electronic circuits, and pieces of metal. The technicians did the translating; few became famous, but many of their names appeared in reports. Without them, Sidewinder would have remained a half-baked idea.

In R&D circles, the distinction between engineer and technician usually hinges on a college degree, typically in engineering. Engineers design things; technicians build or service them. There is a clear line in principle, but China Lake technicians crossed it. They designed gadgets, took out patents, and often assumed important responsibilities. The depression had kept many of them from college, and China Lake gave them a second chance. There were limits, however; the technicians, whatever their proficiency, seldom had a deep knowledge of physics or chemistry. They lacked the formal education of the engineers and physicists, and this created a delicate situation, since the "professional" personnel were dependent on the technicians but also were able to do things the technicians found difficult.

On the Sidewinder project, at least, differences in status were for the most part ignored. Sid Crockett, a technician, came up with the idea for the rollerons, and Earl Donaldson, another technician, was codesigner of the rolleron "pie damper." On the other hand, perfecting the rollerons fell to aerodynamicist Lee Jagiello, whose knowledge of aerodynamics was much more complete than Crockett's.

China Lake put great effort into giving its technicians broad training by allowing them opportunities available nowhere else—specifically, apprenticeships that were the equivalent of two years of college. Their performance was monitored closely; those who did not perform well were not promoted. Frank Knemeyer, head of the Weapons Development Department, was particularly tough in setting standards.

As a result, China Lake technicians were a different breed. They got all-around training and were given responsibilities normally reserved for engineers at other labs. Some of the more talented became inspectors and the best helped develop weapons. Darryl Baxter worked on refrigeration systems ranging in size from 1/6 horsepower to 125 horsepower during his apprenticeship, although most refrigeration engineers never see anything above 25 horsepower. Baxter also did extensive work on the Sidewinder servomechanism controls. He designed, built, and installed parts of Shrike and helped get the missile to the fleet. During AIM-9D development, the Santa Barbara Research Center, a subcontractor, had trouble producing lead selenide detectors. Chuck Smith, head of the development effort, was determined to fix the problem.[14]

> I gave Mr. Jim Foust, my most capable technician, the job of cleaning up the production line at SBRC. For this task, Foust could outperform any of the engineers I had. He set up flow charts of the production line, identified inspection steps, and finally designed fixtures to facilitate various stages in the assembly process. This brought the production under control. . . . We then turned the management back to SBRC. For years after, the engineers and working-level people at SBRC thought Jim Foust could walk on water.[15]

Technician Earl Donaldson built many of physicist Tom Amlie's ideas. Amlie would draw them on a napkin, in classic engineering fashion, and take them to Donaldson for assessment. Donaldson was sometimes skeptical, but usually said, "We'll try it." If it didn't work, Amlie would be unhappy—but he respected Donaldson's ability.

When the Sidewinder project moved to the Weapons Department, however, Donaldson was replaced by an engineer. So he returned to the Aviation Ordnance Department where Amlie assigned him projects equal to his skills. He designed the Scotch-cross gimbal system for SARAH (the radar version of Sidewinder), and later the seeker for the television-guided Walleye. While at China Lake, Donaldson acquired six patents—a respectable number even for a professional engineer.[16]

McLean depended on the technicians, and his favorite was Eddie Kehihian, a toolmaker in the Engineering Department. Ordinarily, designers like Kehihian had only short conversations with the Ph.D.'s, but McLean spent a great deal of time communicating ideas, a practice frowned upon by the Engineering Department, whose rules said that drawings had to go to a planner-estimator who would assign the work to a draftsman. Such a formal system made sense for production in quantity, but McLean disliked it because it lacked the necessary continuity and failed to build corporate memory. McLean, as head of the Aviation Ordnance Department, insisted on having draftsmen and machinists right in his department.[17]

The Learning Organization

The station tracked its successes and failures, and used both to improve whatever it was working on at the moment. The corporate memory system depended on people who had spent years at China Lake and were willing to share their knowledge in many ways: apprenticeships formal and informal, movement of people between projects, internal publications, and courses taught by those with more education or more experience.

The knowledge gained through apprenticeship was important but limited; additional courses often filled in the gaps. China Lake itself offered many courses and the faculty were top notch. Wilcox, for example, was an excellent teacher, and he spent many evenings teaching UCLA extension courses in mathematics, classical mechanics, electricity and magnetism, statistical physics, and quantum mechanics, among others.

Rod McClung taught electronics, presenting all the standard textbook knowledge that one would expect to get in such a course and then illustrating the concepts (and their limitations) with examples from his own very extensive experience. McClung also wrote and published an informal textbook called "First Aid for Projects on the Range: What to Do until the Statistician Comes," which was several times reprinted. Amlie also taught UCLA extension courses and once had William Shockley, coinventor of the transistor, as guest lecturer.[18]

China Lake also had the ability to pick up an idea conceived elsewhere and develop it. "Not invented here" did not often get in the way. China Lakers, McLean included, had their favorite ideas but were more open to outside suggestions than many organizations.

Louis Pasteur, in his inaugural address at the University of Lille, referred to Oersted's discovery of electromagnetism saying that Oersted was fortunate but that "fortune only favors the prepared mind."

The same is true of organizations, which must be ready to accept knowledge, whatever its source. An organization can do this, however, only if it is ready to learn—and this requires a certain strategic flexibility. Entertaining a new thought implies a willingness to act upon it, since all new thoughts imply changes in current thoughts and behavior. China Lake was extremely flexible.[19]

New knowledge arrived in the form of technical literature, discoveries by the Research Department, newly minted Ph.D.'s, and as the result of visits to other projects. Staying current was a constant problem, but China Lake persevered. Competition with other labs was a powerful motive. China Lake couldn't just be good; it had to be *better* than other labs.

[229]

Stay Hungry

McLean was unusual among technical people in having a taste for psychology and sociology, along with a special interest in the nature of high-creativity environments. He had a number of ideas about what kept a research operation productive. He discussed these issues with Tom Milburn, a social scientist at China Lake, and Milburn subsequently studied the management styles of administrators and technical people. (As technical director, McLean got to test many of these ideas.)

McLean believed that creativity was transient—and that it was highest during the period that an individual was in the process of becoming an expert. Creativity declined once the individual became an expert. Bud Sewell discovered this philosophy when McLean suddenly stopped talking to him about warheads:

> [From] 1959 to 1961, Bill brought all his warhead problems to me. In 1961 we got the emergency funds for setting up the conventional weapons program here on the base, and we got a big increase for applied research in warheads and explosives, and the next warhead problem I found Bill had, he didn't bring to me. . . . I felt hurt and [asked] what I had done. "You got funded," he said. "[But] that's what we were trying for!" "Yes, but I don't want you to think that you have a monopoly." He hated monopolies with a passion.
>
> Part of it was almost like the philosophy that [John] Locke had, "You are at the moment what you are in the process of becoming." I

always liked it because it seems so appropriate that it had a rate term tied into it. While you are becoming an expert, you probably are one. If you once declare yourself an expert, you aren't. It's instantaneous. Bill was very much on the side of the becoming. He didn't ask that you be an expert in something, but he did require that you be trying to become expert. See, experts have monopolies.[20]

McLean believed that a research laboratory had only about twenty good years; after that, it was time to pack it up and start over somewhere else. He believed in Rabinow's Law Number 8: "Things that are done illegally are done most efficiently."[21]

He also believed that the key to China Lake's success was an informal network of about two hundred highly capable people who worked at the base. He told Sewell that his major job as technical director was "keeping the department heads from blocking the informal organization from doing its thing." There were people on the base who were listed as experts—and then there were the real experts.[22]

It followed that funds went to whatever projects appeared to the informal leaders to deserve the most support. This often meant that well-funded projects got pillaged to support unfunded but deserving orphans. Early funds to support Walleye, for instance, came out of the Sidewinder budget.[23]

The danger of operating informally is that the person with the most informal power ultimately makes the decision. If the decisions prove successful, everything works. But no one's judgment is infallible, and sub rosa judgments by definition are not subject to public examination. This meant very little support for ideas McLean did not approve. McLean was not always objective. He had pet projects, pet theories, and pet people—and therefore he had blind spots. Those who did not believe in the pet projects often got short shrift. In the fifties Luke Biberman wanted to develop a long-range infrared detection system and hustled up some $250,000 in funds from the Office of Naval Research (ONR). McLean, however, informed ONR that he did not want the project at China Lake and that instead Biberman would work on the Diamondback, a giant version of Sidewinder. Biberman told him that the Diamondback made as much sense as a twelve-pound robin. After that, McLean was cool toward Biberman, who soon left

China Lake for the University of Chicago, where he continued to work on his original project, which eventually developed into the forward-looking infrared system (FLIR).[24]

When he became technical director, McLean got what he had told Sewell was most dangerous: an intellectual monopoly. McLean sometimes found himself at odds with intelligent subordinates because of his ideas about how they should design their systems. While Barney Smith was struggling to keep McLean's hands off of Shrike, Ward was fighting a similar battle over Walleye. Lillian Regelson, associate head of the Weapons Planning Department, said that department heads "never knew when McLean would step in and reverse something." So, many department heads were happy when McLean took no particular interest in what they were doing.[25]

[231]

Others needed little encouragement to operate informally. Frank Knemeyer, head of the Weapons Department, used a systems approach for developing weapons that provided control of project ideas, costs, and interfaces. Out of his department's programs came much of the air-to-ground ordnance used in Vietnam. The system proved effective in developing new ideas and training weapons managers, and it provided accountability—something often neglected by McLean. Knemeyer, however, also knew how to handle funds "creatively" to create seed money for new ideas. To do this he created a "phantom division" that absorbed about 1/7 of the department's overhead. Many of his "seeds" planted by the phantom division grew into big projects.[26]

Techniques and Devices Group

China Lake was shot through with tiny skunk works that become good at doing particular jobs. The Techniques and Devices Group, under the leadership of engineer Fred Davis, is a good example. Created in 1958, the group's basic license was to invent and test devices as needed. When China Lake needed special systems for telemetry and measurement and could not find them on the open market, Davis's group built them.[27]

Although Techniques and Devices formally belonged to the Research Department, it was located two miles away from Michelson Laboratory and reported directly to McLean. McLean visited Davis late each Friday afternoon to talk about what the group was doing. Only four people worked there—Davis, Walt Caffrey, Cass Roquemore, and Lee Humiston—but it had ten benches and a substantial budget.

Fred Davis suggested that one could tell how serious a technical operation was by the ratio of hands-on technicians to engineers. The higher the ratio, the more serious. Davis claimed that as China Lake changed, the ratio got lower. Certainly Techniques and Devices exemplified serious capability. The extra benches allowed the technicians to carry on more than one project at a time. Large, unionized laboratories at many defense contractors did not permit the engineers to touch tools. At China Lake, however, engineers and technicians worked side-by-side at their own benches.

Techniques and Devices had materialized to fulfill a need and was used to produce key items. It lasted nine years and was highly creative throughout. After McLean left in 1967, Hugh Hunter, head of Research, took advantage of Fred Davis's absence in Vietnam and closed it. Henry Swift and Larry Nichols tried to reestablish the group later, but in vain.

Technological Maestros

Arthur Squires, veteran of OSRD, has argued that successful projects have "technological maestros"—people with technical virtuosity, high standards, a high energy level, and a formidable attention span.[28]

There are different kinds of maestros, and China Lake had many. Some were technical virtuosos who put forth novel ideas or got things to work; McLean certainly fit into this category, as did Jagiello, Howard Wilcox, Woodworth, and Crawford. Others were implementers and problem solvers, such as Chuck Smith and Burrell Hays, who could accomplish the seemingly impossible in terms of organization and management. Still others, such as Hack Wilson, Ward, Knemeyer, and Doug Wilcox, knew how to create an organization that would facilitate high creativity.

[232]

Here is Crawford, a technical maestro, reflecting on Ward as a leader:

Well, I think that Newt Ward was really the top guy . . . hands-off and yet hands-on. He's hands-off in the sense that he doesn't direct what you're doing, but he's very much hands-on in the sense that he knows what you're doing because he's taking the time to come around and keep tabs on that, not in the sense of checking up on you, but because he really wants to understand what's going on. So you looked on him more as somebody who was trying to help you than somebody directing you.[29]

Here is Ernie Cozzens, associate head of engineering, describing Ward: "Newt Ward knew every person on this center by their first name. I often marveled at how that man could know everybody by their first name. Dr. Ward spent a great deal of time walking around into the spaces. And he didn't restrict himself to his own department; he'd walk around everyplace, and he talked to people, and he made you feel good."[30] This kind of supportive, hands-on leadership, well known in the industrial world as "management by walking around," characterized much of China Lake's practice.

[233]

Mickie Benton noted the importance of managers who were technically sound:

> One of the things that I observed in the early days is that we had management that was strong technically. The ones that we've been talking about, those were our managers . . . but they were also inventors. . . . They could recognize ideas that had been presented and then they could evaluate them on a good sound technical basis, rather than on some intuition as too often turns out to be the case in some environments today.[31]

Chuck Smith, who headed the Sidewinder project during AIM-9D, used weekends to solve the problems his team had accumulated during the week. Like McLean, Smith had an intuitive grasp of the technology and its operations. Usually his interventions got his group back on track.[32]

China Lake's leaders were recognized for their ability to ask the key questions. Wilcox recalls a China Lake commander, Capt. Walter Vieweg, coming down to the "sweat-stained spaces" where Sidewinder was under development:

> I told him that I expected to be successful within a year. "Have you solved your problems?" "No," I said, "[not] yet." "Well, take this gyro

wobble problem, for example. Have you solved that?" And I said, "No, not yet." "I see," he said, "you're going to be successful within a year, but you haven't solved your problems yet. Is this what I am being told?" Well, my estimate of Captain Vieweg moved up many notches when he said that.[33]

Ernie Cozzens recalled his experience with Francis Carlisle, another maestro:

Francis wasn't a hard-charger; he sat most of the day with his feet up on the desk, and he didn't talk very well; he didn't stutter, but he would say "uh—uh—uh." Let me tell you something, today we've got a deputy technical director, we've got almost the entire management of this [engineering] department with the exception of Jack Russell, we've got division heads all over the center, who worked for this guy, and what he did was instilled in everybody a desire to learn, an inquisitive mind. He could ask question [after] question, but he didn't ask because he wanted to know: He asked to find out if they knew, and if they didn't know, they felt bad because they couldn't give him an answer. And as a consequence they learned; they went to school.[34]

As a division head in the Engineering Department, Carlisle carefully trained his production engineers in design. Only years later did Burrell Hays, for instance, realize that Carlisle had carefully nurtured his design abilities so that when he became involved with production, he would understand designers' problems.[35]

McLean As Maestro

In evaluating McLean, it is useful to consider an example from another context. Describing Enrico Fermi's style in leading the group constructing the first nuclear reactor, physicist R. R. Wilson said:

This was intrinsically team research from the beginning. A group of nuclear physicists, dominated by the force of Enrico Fermi's genius, did essentially what he wanted them to do. I do not remember Fermi ever issuing orders about exactly what should be done or specifying who should do it. In a formal sense, he was not even in charge of the work. However, because of his brilliant theoretical analysis of our problems it simply seemed self-evident what was to be done next and who was to do it.[36]

"Dominated by the force of [McLean's] genius" would certainly be an accurate description of the Sidewinder project. J. Raymond Schreiber, in the Target Tracking Branch, commented, "I think we were all caught up in Bill's dynamic personality, if you want to call it that, and how he operated."[37] Howie Wilcox, defending McLean's style of project management, stated:

> What do the people want to do? My argument was that people want to do anything that a forceful project manager wants them to do. They want to feel part of the thing, to feel needed and all one needs to be is a forceful leader and say, "This is what I want you to do" and they would be glad to do it. Most people don't want to make up their mind about large, abstract issues. They want the thing laid out in such a way that it's clear. . . . And they are actually willing to suffer a fair amount of discomfort to be part of a going operation and feel like they are making a contribution.[38]

McLean did not dominate through force of personality—Henry Swift confessed that it was difficult to get McLean to give him any direction— but through force of ideas. Of course, the relationship between McLean and his coworkers was symbiotic. McLean would not have been able to operate successfully without a supporting cadre of others who were willing to undertake the many tasks for which he showed no inclination.

"Cosmic management," as Frank Knemeyer called the base's top brass, was expected to support such creative activities. Usually it did. McLean certainly did so as the station's technical director from 1954 until 1967. So did Hack Wilson, the associate technical director during this time, and so did many of the other technical directors. Commanders were a more mixed bag. Some, such as Captains Stroop and Ashworth, were highly respected. Others proved disappointing. Some who came later were actively disliked.

The Emotional Infrastructure

One obvious question—in this highly masculine society—is how the women fared. The fourteen-hour days might be good for male bonding, but what about marriages and children? Some women were secretaries, engineers, or even Ph.D.'s who worked at Michelson or elsewhere on the base, but most were spouses whose major responsibility was homemaking and looking after children. Nevertheless, China Lake was remarkable in that many women filled technical positions. Lillian Regelson estimated that when she was hired China Lake had about half the female technical professionals employed by the United States government; few of them had children.[39]

Families at China Lake thrived in the desert far from relatives and civilization once they adapted to the local culture. Since the men were usually at work, even in the evening, the wives had to be independent. Still, families and children were part of a tightly knit system, and all members of the community tended to perceive a common bond. Children had freedom and learned self-sufficiency at an early age. Since there are hazards to living on a desert military weapons base, they had to be careful.

Interesting people and plenty of social life on a small base precluded the isolation experienced by suburban housewives of the 1950s. Chances are that the next-door neighbor's husband was doing something similar to one's own husband, and that helped.

And while work penetrated the family circle, it also brought neighbors and like-minded couples together. Sports and athletic activities of all kinds were popular. There were clubs for everything; Vice Admiral Moran estimates that 180 clubs existed in the early 1950s.[40]

There were challenges, but there was a supportive infrastructure. For many years the maestro of *this* system was LaV McLean. She appears on the cover of the third volume of China Lake's history; this is not just by courtesy. LaV was what sociologists call a "socioemotional leader."[41]

She got to know most of the children through her work as a gym [236] teacher at Murray Junior High School. Through the children, LaV had her finger on the pulse of the entire base. She understood peoples' needs and was responsive to them far beyond any ordinary expectation. Once when a family had a car accident near the base, LaV found that they had no way to get to the hospital to visit a family member there. So she loaned them her car until the injured person was released from the hospital. Years after the event, she was still in contact with the family. Whenever a new engineer joined the base as a junior professional, LaV got to know him and his family. She gave a party for all the new engineers at the Officers Club. She also kept the scientists and the military together, always a problem given the two cultures. "LaV was our Number One ambassador as far as keeping the military and the civilians as one, keeping a good balance. It took a lot of doing, but she was able to. That was very important in those early years, but it couldn't continue . . . as the base got so large. But she did an extraordinary job, and I think that was why the base was such a success as far as the social life. Her parties always involved as many military as civilians."[42]

She also was an important asset for China Lake's external relations. Her easy confidence, phenomenal memory, and complete inability to be intimidated endeared her to bureaucrat and scientist alike. At parties for visiting officials, chances are that Bill McLean would be off in a corner talking about guided missiles or submarines. LaV, however, would be keeping the visiting dignitaries entertained. She knew how to make a party come alive, both for the junior engineer and for the senior admiral. China Lake would not have been what it was without LaV.

The McLeans would alternately dazzle visitors with science and warmth. After Bill had fired up their imaginations, LaV would introduce them to the social life on the base and to the beauty of the desert: "We'd take them up to Coso Springs or out on the ranges. A lot of people didn't have any idea what the desert was like, and we thought the desert was a beautiful place; it's got a beauty all its own, a quietness. It was so good for these people to get out and away from the hustle and bustle. They would loosen up a bit and enjoy each other." Visitors often stayed with the McLeans, and discussions often took place around the McLean breakfast table or in the living room of their home.[43]

Contacts like these made it easy for McLean to call Adm. Arleigh Burke, while he was chief of Naval Operations, or Jim Wakelin, assistant secretary of the navy for R&D, and get problems solved on a one-to-one basis. When McLean retired, such contacts were more difficult.[44]

LaV was not the only person, however, who acted as an unofficial host to visiting dignitaries. McLean's office manager, Nona Turner, also served in this capacity. Nona knew a great deal about Washing-

LaV McLean. Courtesy of China Lake *Rocketeer*

ton and about the base. She had come to China Lake as a WAVE, and she understood administrative requirements—as well as Bill McLean's eagerness to avoid them.

When McLean returned from a trip, she would sit him down, find out what commitments he had extracted from the bureaucrats, and then report to the department heads what funds and contracts Bill had gotten for them, and what he had committed them to do. Under McLean, Nona "ran the base," along with her counterparts Dottie Dunn in Hack Wilson's office and Jo Ritchie in the commander's office.[45]

Visiting navy brass also knew they were welcome at the Turners' house after an evening of formalities. Nona and her husband, Harold, provided a place where they could relax. The access this provided

[238]

to senior military leaders was not abused, but it certainly did not hurt the base.

The kind of entrée this familiarity provided in Washington is impressive. In the late 1960s Bud Sewell, Nona, and Dottie were in Washington representing the Center for the Vietnam Laboratory Assistance Program. Nona by then worked for Sewell, and Dottie was representing Hack Wilson, acting technical director. The women had come along so they could learn the faces that went with the names on the correspondence they handled.

[239]

After a long day of meetings, Sewell went out to dinner with the just-returned chief of the Navy Research and Development Unit, a navy commander. Nona and Dottie, meanwhile, went out to dinner on the personal invitation of Adm. Thomas Moorer, Vice Adm. Thomas Connolly, and Vice Adm. Levering Smith—all China Lake alumni— who were respectively the chairman of the joint chiefs of staff, the deputy chief of naval operations for air, and the head of the Polaris and Poseidon programs.[46]

For many years, China Lake was able to count on high-level support from officers who had worked on the base and understood its culture; many held senior positions with the bureaus (later systems commands) that controlled money and projects. This support was critical in funneling projects to China Lake, protecting it from interference, and ensuring recognition of its value. When the carrier admirals were replaced by flag officers from the nuclear navy, that is, submariners, China Lake's support dwindled.

The isolation of the desert, the mixture of civilians and military living together, and some very fortunate personality choices created a unique environment. Little seemed to stand in their way except the laws of nature and bureaucratic pressure from Washington. They worked within the laws of nature and often out-maneuvered the bureaucrats. Part of the mixture was luck; but fortune, as Pasteur said, favors only the prepared mind.

[18]

Building Ideas at China Lake

They thought of us as the Huns descending on Washington out of the west to run off with all the plums. I think there was a certain "robber baron" character to the way [China Lake] operated, because our philosophy basically was: if you can run faster than the next guy, you get the job.

Howie Wilcox

Designer Weapons

The end of World War II marked a turning point in the relationship between the United States and its R&D community, which had become a political force. *Making* weapons has been big business for centuries. But after the war *developing* weapons became big business as well. To cope with this transformation, government changed. During the war OSRD had concentrated on developing weapons; whatever needed to be done was done, and done quickly; it was a skunk works on a national scale. But after the war, the bureaucrats took over. Getting weapons that work has gotten harder.[1] Here's why:

1. Since weapons design decisions confer big money and power, they cannot escape political pressures. Nor is it likely that any number of rules or penalties can eliminate corruption. Contracts provide jobs

and political pork; a military service gains power over its rivals when it develops a weapons system; minority businesses are aided by making sure that contracts are distributed broadly; system failures provide politicians with ammunition to embarrass their opponents. Repeated attempts at reform have run aground not only on the realities of politics but also the complexities of weapons R&D.[2]

2. Procurement scandals have led to ever more rigid checks and balances. By the 1980s the Pentagon's book of purchasing rules had grown to seventy-five hundred pages, with another thirty thousand pages of accompanying policy guidance. Yet the rules often fail to restrain political juggernauts, heavy-handed sponsorship, rigged tests, and failed military oversight. Classic examples of such failures include the M-16 rifle, the C-5A Galaxy transport, and the Division Air Defense (DIVAD) antiaircraft gun.[3]

[241]

In spite of the controls, corruption runs rampant while reformers and whistle-blowers receive severe punishment. Ironically, it appears that during World War II, when deals were often sealed with a handshake or a telephone call, business in the defense area was far more honest and involved a fraction of the waste present in today's highly legalistic environment.[4]

3. Red tape imposes substantial costs in time and money, and failed projects. Rules designed to prevent cost overruns keep many good projects from being developed. In addition to thinking up the weapon and getting it developed, R&D teams must use their ingenuity to follow or to circumvent bureaucratic rules. The reader will have detected a substantial amount of rule-breaking in the Sidewinder program, but this rule-breaking was to allow the team to implement its vision. A general consensus among people consulted while writing this book was that the Sidewinder development process could not be repeated today.

Efforts to get around the rules sometimes founder. Some cynics believe that "black" programs are created as much to avoid bureaucratic scrutiny as to improve security. While successes like the F-117A Stealth fighter have emerged from the process, other systems needed more scrutiny, notably the air force's B-1 and B-2 bombers, and the navy's A-12 attack plane.[5]

4. It is harder to produce effective new systems. The weapons we do get often work badly. Even when the systems work, they cost too much, which means the armed forces must make do with fewer of them. Some supporters believe that high performance can compensate for reduced numbers. Others strongly disagree, noting with concern how critical a single system's success or failure has become—and that one airplane, one ship, or one tank can still be in only one place at a time.[6]

Useful weapons do get produced. Overall, however, the United States spends too much and gets too little, especially where big-ticket items are concerned. And these "weapons systems" take the majority of defense procurement dollars. U.S. military aircraft, for instance, often have capabilities no better than foreign systems, in spite of higher development costs.[7]

A China Lake Way?

Over the years, many studies have suggested better ways—skunk works, for example. China Lake's Sidewinder development process shows skunk works methods operating on a grand scale. Such organizations are not a panacea, however: Boeing's 777 probably could not have been built using skunk works methods.[8]

When Sidewinder was developed, the rules were few and flexible. China Lake's success rested on the vision of its founders, who continued the highly creative wartime subculture of OSRD—a major achievement in itself. The main lines of this organizational culture were spelled out in China Lake's principles of operation, and China Lake's habits of thought and action made it thrive. The resulting mix of written and unwritten practices provided an organizational dynamic that allowed China Lake to succeed at tasks other laboratories considered impossible. How did it work?

The Creative Dialogue

The navy directed China Lake to research, develop, and test new weapons. In principle, its tinkering was to take place in response to formally approved operational requirements; the navy would decide what it wanted; China Lake would develop it; and industry would manufacture it. In many cases, however, China Lake decided on its own projects. It studied a navy problem, developed an understanding, created the system or the technological fix, and sold the idea back to the navy. The selling often involved a dialogue with a network of supporters, funders, and users. The process took place with guided missiles such as Sidewinder, Walleye, and Shrike. Through such dialogues, China Lake evolved its own mission.

The key to maintaining the effective dialogues was a system built on five key principles:

[243]

Get the right people.

Give them the right challenges.

Give them the proper tools.

Put as few barriers in their way as possible.

Hold them accountable for success; ignore failures.[9]

What was China Lake's real role? Many, including McLean, thought it was to define what concepts were needed and make the first moves toward their solution. With guided missiles, the laboratory had started out doing only testing and evaluation, the culmination of the R&D process. But China Lakers showed themselves adept at solving technical problems on all phases of a project. The key was allowing the natural visionaries to formulate their visions and then organize teams to implement them. Don Moore, an engineer, commented: "I think the important part of the culture was to be able to draw on people who had little pockets of expertise of one type or another. There were enough people around that saw the big picture. . . . Sooner or later a concept would evolve that looked viable to everybody and we would charge off and do it."[10]

Being able to choose its own problems, "the things that really needed to be done," as well as those passed down from above, was essential to make China Lake a successful laboratory.[11] China Lake's first technical director, L. T. E. Thompson, thought an engineer should spend 20 percent of his time working on projects that he thought were most important. Five percent of the institution's funds also were reserved for such personal projects.[12] Not all inventions were responses. Sometimes, as Don Moore acknowledged, a person evolved a solution and then looked for a problem to which it could be applied.

Since the military procurement system did not envision this license to think, struggles over the mission took place inside China Lake as well as outside. Even at China Lake, some highly gifted people felt that the center's strength was its ability to *test and improve* systems originated elsewhere. They saw China Lake as a center of expertise. Others felt the emphasis ought to be working on whole weapons systems not subsystems. Wilcox, taking over as head of the Rocket Development Department in 1956, announced that "China Lake was not going to do any more 'cats and dogs,'" the subsystems that many groups considered their stock in trade, but instead would concentrate on developing whole weapons systems. Many opposed Wilcox's point of view. Certainly outside China Lake there were many who felt China Lake should not be originating weapons at all, but instead should be coordinating them through contractors.[13]

"Do the Right Thing"

The essence of the China Lake approach could be put into four words: "Do the right thing." This is supremely important in understanding the value of environments like China Lake. But why wouldn't a laboratory do the right thing—especially if the alternative is doing something less than the best? There are a variety of reasons:

The organization is trying to do what the buyer has requested, even though it is not a particularly good idea.

Schedules, rules, or requirements force a narrowing of options leading to the wrong choice.

The research operation is micromanaged by the client, stifling creativity.

The research is carried out on a nine-to-five basis. *Sidewinder never would have been produced by people working normal hours.*

China Lake's isolation and success enabled the station to avoid most of these pitfalls. The navy tolerated China Lake's behavior because it got things done. But the social contract formed between the navy and China Lake required that the laboratory maintain its edge. Underlying its success were certain preconditions for doing the right thing.

Getting the right people in the first place paid off. There was a fair consensus among those interviewed about who was good at what. Those running the show also believed that native ability required nurturing through training and, if necessary, through more formal education. Virtually everyone interviewed was sensitive to the process of technological apprenticeship, by which judgment was learned. A high degree of sensitivity to where people were in the learning process was used to guide education. Most people could trace not only their own career but also those of others through a series of projects and thus implicitly through a series of apprenticeships. The weapon designers had to work effectively as part of a team.

Another precondition was that rules should take a backseat to intelligent action. Because rules cannot be ignored completely, they were bent—a China Lake habit. While common in small-scale skunk works, this ethic prevailed also at effective larger laboratories like China Lake. Under McLean, paperwork was used to document what was important, not to protect against recriminations. Individuals who did not grasp what could and could not be done sometimes felt frustrated. Not everyone was comfortable with the level of bureaucratically deviant activity that took place: those who weren't migrated to analytical and research functions rather than working directly on development.[14]

A final requirement was that solutions had to respond to the users' needs. Technical virtuosity took second place to user-friendliness. This implied close interaction with potential users over design features not only on paper but also in the air, at sea, or in the field. Systems had to be simple, reliable, and effective. China Lake emphasized uncovering key problems, which were then attacked using the station's formidable array of talent, techniques, and corporate resources.

Stretching the Rules

China Lake operated as if there were a war on. Ernie Cozzens, associate head of the Engineering Department, in praising Newt Ward noted: "He was the guy who believed to get the job done sometimes you had to bend the rules and bend them pretty bad. I was working at the Sight Lab one time when they dragged a loaded rocket motor of some kind in there. They wanted a bracket on it. Here's the guy drilling holes in the thing, and here's the Center Safety Committee having a safety inspection in front of the building, and Newt's just telling [them] how great it is, everything's safe."[15]

Rod McClung recalled an incident involving McLean:

You had to stand in line at shop stores to draw out electronic components or anything, and Dr. McLean insisted that they open it up supermarket-style so you could just go through and then check the stuff out. The Supply Officer said, "But Doctor, they'll steal us blind if they can just go in and pick the stuff up." And Dr. McLean looked at him and said, "If the people that are smart enough to develop these missiles aren't smart enough to steal you blind any time they feel like it, I don't want 'em."[16]

Capt. P. D. Stroop commanded China Lake for only ten months, but he got a lot done. His message to the troops: "Do what you see as the best thing to do and I'll cover you from the rear."[17] A launcher to test Lark missiles was built from funds intended for a chapel; the Lark funds had gotten delayed, so money was borrowed from the chapel account. The Lark funds never arrived, so China Lake went without a chapel for quite some time. McClung said he would go to his grave with some secrets regarding the worst sins committed on projects he had worked on, all in the interest of getting the job done.[18]

"All money is green and it all comes from the taxpayer," was a common China Lake phrase. Hack Wilson, Bill McLean's associate technical director, gave weapons planner Lillian Regelson some valuable lessons about operating in the government system. "He told me that he was very careful to be sure he understood how far he could stretch the rules about shifting money around without being vulnerable to being sent to prison. Short of that, he would say, 'All they can do is give me ten lashes with a wet noodle.'"[19]

Capt. Frank Ault, no stranger to getting things done sub rosa, found that China Lake set new standards for subterfuge and reprogramming funds. In the late 1950s, when Ault was working in Pentagon R&D funding, he found that China Lake reprogrammed funds scant hours after they had ostensibly agreed on priorities. He said he had the feeling the paperwork had already been prepared during the meeting, and the minute he left, calls would go out from the base to execute whatever prearranged plan China Lake had already cooked up. Ault once threatened to physically abuse Howie Wilcox, who had gone over Ault's head to get additional funds for the Notsnik satellite project.[20]

A Community of Good Judgment

China Lake culture was transmitted by learning on the job such principles as solving the right problem, designing for rapid feedback [247] from tests, or finding out what the user really wanted. Little respect was accorded cosmic managers who consciously tried to eliminate the informal culture, as did Rear Admiral Freeman. But there also was frustration with those who inadvertently harmed the culture by flagrantly breaking the rules and endangering the station's relationship with the navy or other laboratories, which invited jealousy and retribution.[21]

Making mistakes with hardware was essential to this learning process. Personal experience teaches lessons hard to forget, as Jack Crawford observed concerning a Terasca rocket launch:

> The timer controlling the second- and third-stage firing was sent a commitment signal about two seconds before launch, and a return signal verified the counter was running. On one launch, a hold-fire signal was received one second before launch. That stopped firing of the first stage but the timer controlling the second and third stages was committed. Right on time, at 20-some seconds, the second and third stages took off leaving the first stage behind on the launcher. In retrospect the logic flaw was obvious, but no one had caught it previously. As a personal aside, I think the impact of going through some of these real hardware mishaps is missing in the computer-simulation world. When you are in a firing bunker and a rocket takes off unexpectedly, you are a changed person. You realize the importance of considering all the possible scenarios even if some of them "can't happen."[22]

In time, however, this hands-on experience became diluted. Changes in the organization's mission made such hands-on experiments fewer and further between. Simulations replaced physical tests. Many felt that China Lake's ability to act as an intelligent judge of others' systems also declined.

"A Culture of Pure Work"

Many people, McLean especially, noted the advantages of personnel living on the base in terms of commitment to work and continuous problem solving. Work was carried out whenever it could be: in the labs, on the test ranges, over cocktails, at the Officers Club, and over the back fence. Work did not stop when people went home; it simply shifted location. Someone with an idea could call McLean at any time during the evening or on weekends; the caller would be invited over to his house to discuss his idea. Wilcox was perfectly capable of calling a meeting on Sunday morning.[23]

It was, in the fine phrase of Teresa Amabile, "a culture of pure work." The atmosphere encouraged a huge attention span. Bits and pieces of one project were always around ready to be connected to another project if they were relevant. There was always a loosely focused interest for important projects in general. The entire place was serendipitous.[24]

The intellectual atmosphere at China Lake was supersaturated and only waited for some catalyst to bring down a rain of ideas. Military needs were seen as stimuli to the group's creativity, and the creativity was high pressure and continuous. It was easy to be caught up in the flow of daily innovation. The ability to proceed directly from any point of the base to the laboratory meant that ideas got tried quickly. Many interviewees connected the decline of China Lake with the growth of Ridgecrest, the town surrounding the base. When civilians moved off the base and into town, there was less inclination to go back to the laboratory after working hours, a greater separation of work and play. Comparing China Lake with another naval facility in San Diego, Ken Powers suggested explicitly that creativity declined when people moved off base.

[248]

[At China Lake] we had good information on what was going on not only on our own programs, but also on programs that might in the future contribute to our programs. Our morale was high; we knew the people; we knew where to go to get help; we had more of a feeling for the total program; the informal communication was fantastic. When I left NWC [China Lake] in 1972, I went to McLean's new undersea laboratory in San Diego, and the change was phenomenal. I think once in the first three months after I got there I saw one person out-side of the work environment.[25]

Efforts to develop projects were continuous, free-wheeling, and knew few boundaries, including those of the base itself, which may explain one of the puzzling features of this skunk works environment: the low burnout rate of technical personnel. While projects required enormous amounts of effort—sometimes twelve hours a day, seven days a week, year after year—the amount of cynicism at China Lake was low.

[249]

At least, this is the way they remembered it. Several factors were involved in the low burnout rate, but the largest was *empowerment.* Because people got to work on projects they thought were worth-while, the impersonality of many big technical projects was absent. A sense of personal power and responsibility made for high morale.

Bill McLean and Sidewinder developers in the McLeans' home. *Left to right:* Walt LaBerge, Howard Wilcox, Bill McLean, LaV McLean, Newt Ward. Courtesy of LaVerne McLean

Awarding credit was important. Often enormous effort would be put into a project because it was one's own project. The feeling that the project was "my baby" reached from the ranges right down to the model shop. According to Barney Smith: "We sidestepped the planners and estimators and went right to the person who was actually going to build the machine. When you take the suggestions of the machinist, give him credit, have camaraderie with him, and then he sees his device being tested up in the air—he is on your team." At China Lake, there was a sense that those who got the rewards deserved them. Deviations from this principle stood out.[26]

It was not all sweetness and light. Barney Smith, inventor of the rocket-assisted torpedo, and a highly capable project director, became head of the Weapons Department following Wilcox. Yet Smith left China Lake in 1959 worn out. He had failed to get Notsnik funded, struggling with the "impossible conditions" imposed on him by Herbert York, director of the Advanced Research Projects Agency and Wilcox's overoptimistic budgeting.[27]

Smith also had struggled to protect his subordinate Frank Cartwright, a gifted division head with a short fuse, and he struggled with McLean, who thought he had a better way to design the Shrike antiradar missile than the way Jagiello was doing it. Smith felt burned out but hardly ready to retire. He took a one-year sabbatical at the Naval War College, in a think tank run by Adm. Edwin Hooper. Then he was recruited by former China Lake commander Adm. Dick Ashworth, who was head of R&D for the Bureau of Naval Weapons; Smith became chief engineer for Ashworth's division. Then he spent his last ten years in navy service as technical director of the Naval Weapons Laboratory at Dahlgren, Virginia. China Lake could prove too much even for an energetic manager like Smith.[28]

[250] ## China Lake and Washington

The navy was lucky to have a China Lake. Secretary of Defense Robert McNamara remarked that 75 percent of the air-launched ordnance used in Vietnam originated there. Yet this high performance, praised in speeches, created ambivalent feelings within the navy.[29]

China Lake freelanced by spending a great deal of time trying to figure out what the fleet needed—and then letting the fleet know that its needs could be met, if only China Lake could get the funds to do the project. Washington often was in the dark until the last minute. During the Vietnam War, navy F8U photo reconnaissance squadrons wanted radar-warning receivers to find out when they were being tracked. Engineer Gerry O. Miller at China Lake had such a detector, and the squadron commanders told him they wanted to try them. Accordingly, the detectors were soon installed in a few planes. In Vietnam the devices worked so well the entire squadron wanted them. China Lake told the squadron it would have to get naval air to approve the installation. When NavAir officials found out what had been done, they wanted to know what license China Lake had to modify operational aircraft. Thus while China Lake had good rapport with the fleet, its relations with Washington often left much to be desired.[30]

[251]

Occasionally, China Lake would find a powerful civilian whose agenda matched its own. John Rexroth was such a person—a senior, experienced technologist at NavAir who worked with China Lake on systems R&D, including quality and production problems. Rexroth, too, put the fleet first, and he helped many key projects. Others, such as Assistant Secretary of the Navy for R&D David Potter, thought China Lake was arrogant and needed taming; Elizabeth Beggs, the senior NavAir civilian in charge of guidance, made what China Lakers considered capricious decisions and canceled many of their key projects. And then there were government officials such as Melvyn Paisley, assistant secretary of defense for R&D. Paisley, who later went to prison for taking bribes, wanted China Lake silenced because it interfered with his personal projects.[31]

When Paisley insisted that China Lake give the Tacit Rainbow unmanned aerial vehicle a positive review, Burrell Hays (technical director, 1982–86) refused to bow to pressure and eventually resigned rather than give in. China Lake, in a massive show of support for Hays, rented the biggest facility in Ridgecrest and two thousand people showed up to honor him.[32]

Cinderella

Success breeds jealousy. In their competition with other laboratories, China Lakers sometimes acted like pirates. Again and again, they succeeded at projects considered completely unworkable or impossible by other laboratories. Jack Crawford's assistance to the Condor project at North American Rockwell is a good example. While Crawford's (and later Nadim Totah's) timely assistance improved the weapons system, it also caused strain with Rockwell.

China Lake was perfectly capable of taking over a project and getting it to work in a very short time. The antisubmarine rocket (ASROC) fits this definition. Barney Smith managed to outtalk the competition in Washington and convinced the Pentagon the system should be designed at China Lake. On a Friday, Smith told the navy that China Lake could develop the thrust cutoff device necessary for the success of its idea. On the following Monday, he showed up with the thrust cutoff device in hand. It had been developed over the weekend. "We used to pride ourselves at China Lake [on being] able to scrounge enough spare parts to build one of anything on any given weekend."[33]

China Lake prided itself on being able to beat others to the punch, and this "we can do anything" attitude incensed its rivals. Even McLean became sensitive to the hostility building up. He actually terminated the superior Marlin submarine bombardment weapon in deference to another project, SUBROC, under development by the Naval Ordnance Laboratory at White Oak, Maryland, "so that White Oak would have something to work on." McLean felt that if they kept getting rivals' projects, there would be an explosion. The explosion happened anyway.[34]

China Lake's Shrike missile system proved to be a viable antiradar weapon. In the late 1950s, however, elsewhere in the navy, the Crossbow antiradar missile also was being advanced for the same mission. The nearby Corona Naval Ordnance Laboratory was developing it under Fred Alpers; as Shrike matured, Crossbow's days were numbered. Finally, it was canceled. Avocet, another Corona project, also was canceled—to be replaced later by the similar, but more advanced, China Lake Walleye. While there were many differences between the two conflicts, such conflicts were often resolved in favor of China Lake.

Envy grew when China Lake succeeded with projects stolen from other laboratories; its arrogance was legendary. Most believed that China Lake was going to do what it wanted to do, no matter what it was told to do or budgeted for. Engineer Bob Sizemore remembers being shown an organization chart in Washington that had a penciled-in box labeled "NOTS [China Lake]" above the chief of the Bureau of Ordnance.[35]

In the fairy tale, Cinderella meets Prince Charming and lives happily ever after. In real life Cinderellas are often held back by the powerful and jealous older sisters. And when China Lake lost Bill McLean—its very own Prince Charming—the older sisters closed in for the kill.

[253]

[19]

The Past and Future Laboratory

I think that it [China Lake] has gone back almost to an inoperable status.

Bill McLean, 1975

At one time you could walk down the first and second floor of the lab—and it was *lab* space. It's now office space. There is a considerable difference.

Woody Woodworth, 1993

Agile Systems

In late 1994 I spoke with Capt. Frank Ault, father of Top Gun. Ault was both an admirer and a critic of China Lake. He thought the station had become too wedded to Sidewinder and that it had failed to carry through on his report's most important recommendation: To develop a genuine dogfight missile with thrust-vector control.[1]

China Lake, under Frank Cartwright's leadership, had in fact tried to develop such a missile—Agile—which reached the prototype stage. In the early 1970s Rear Adm. Bill Moran, then commander of China Lake, and Frank Cartwright, then in charge of the program, watched the prototype proceed faultlessly through a complex evolution—up, turning 180 degrees, down, and so forth. Both men were impressed. For a variety of reasons, however, Agile never left the laboratory.[2]

In 1994, noting that Sidewinders and Sparrows were still the standard air-to-air missiles, Ault observed that "China Lake culture is still with us." The United States is still using fins rather than thrust-vectoring to control missiles. It is true that the AIM-9X, still under development, will move the United States into thrust-vectoring technology, but China Lake is not developing it. Meanwhile, the Russians have fielded and exported an off-boresight, helmet-sighted, air-to-air missile with thrust-vector control (the R-73 Vympel) that works.[3]

[255]

What has happened? Did China Lake get too set in its ways? McLean had worried that China Lake was losing its creativity; that may be one reason why he left. He thought creative organizations were subject to mental hardening of the arteries and were good for twenty years—at best—after which it was time to break them up and start again. This idea came from his theory of expertise: You are an expert only when you are becoming one; when you finally *become* an expert, you already are losing your abilities.[4]

China Lake has published a fine chart of its history and technical accomplishments—a visual display of weapons and, to a lesser extent, of component and software/process improvements. At first glance, the chart does not reveal any hardening of the arteries. But closer study suggests that innovation was strongest in the years before McLean's departure in 1967. Insiders, especially the alumni of the 1950s and 1960s, feel that China Lake has lost its creative power. Many believe that when McLean left and the name changed from NOTS to Naval Weapons Center, the station's character changed also.

China Lake was very creative as part of Caltech during World War II and in the 1950s when Sidewinder, Shrike, Notsnik, and Walleye, all "firsts" in their respective areas, were invented. The Pasadena Annex developed ASROC during this same period. During the Vietnam War China Lake was in its element, constantly turning out new systems. By the 1980s, however, China Lake had shifted from designing new systems to doing updates, software, and support. Technically, it remained outstanding; it was excellent at fixing missiles others designed, and it still had many a skunk works team. It could still get things done in record time; it had outstanding test facilities; and it could still design. But it innovated at a lower level. There were no more Sidewinders.

Moving civilians off the base and into the growing community of Ridgecrest changed things. When civilians and military mingled, the station had served as a pressure cooker for ideas. When the civilians left, some of the pressure went out of the cooker. Peggy Rogers, a department head whose judgment was widely respected, thought that closing the cafeteria was another major blow to creativity, since it, too, was a place where brainstorming and similar creative activities often took place.

Probably the most critical failure, however, was the failure to maintain the "community of good judgment." This became evident to many in the quality of appointments at all levels in the organization. Without talented leaders, the troops began to lose the will to think, thus illustrating another of Rabinow's laws: "When the boss is a dope, everyone under him is a dope or soon will be." A high-empowerment culture like that of China Lake cannot afford carelessness when selecting those who fill its key positions. Dick Schmitt, a former China Lake missile designer, put it this way: "The way China Lake used to do development [was] . . . small group, high pressure, and the authority to change it and make it go the way you want to make it go. The problem was that China Lake was not able to control that and *keep* it. Maybe there's even a sense of this arrogance or a sense of not really understanding what they really had. Not keeping the group of people there, and not keeping the management discipline that ought to be involved."[5]

Carpets and Computers Replace Lab Equipment

C. Northcote Parkinson, in a satirical book, reflected that there is an inverse relationship between the luxury of a laboratory and its productivity.[6] Leroy Riggs, former acting technical director, reflected on the changes at China Lake over nearly forty years:

> When you look at the things that made it happen, as important as say Bill [McLean] was, it was important that it was a terrible place to live, with lousy government quarters, using secondhand machines; all your old vehicles were World War II leftover trucks, that instead of cannibalizing they sent out there.

I remember when I went to Mike Lab, it was only eight months old and it was a *laboratory.* I mean you could walk down every one of those wings, and you would have high-pressure gas, air, vacuum, several kinds of power, and battleship linoleum floor. I think Hack [Wilson] had some carpeting up there, and the department heads had carpeting in their personal offices. You go up there now, and *mail clerks* have carpets. And you walk down any of those wings, and every wing, except for Chem wing and the Physics wing, which are wings four and six, every single one of them has had all the laboratory equipment taken out. There is no fancy power, no gas, no vacuum. They are offices, carpeted offices, with people shuffling papers.

[257]

Now there are a few little teeny enclaves where guys hide, and you probably won't find them on the code chart. But in today's system, it's hard to bust out of that 6.1, 6.2 [exploratory funding] kind of area. There's no way they can get it to a point where they can literally go to Washington, and sell the project as a new project. They'll say, "Gee, what requirement did this come from?" There's a lot of hobby-shopping going on, but I don't think it's going to turn out to be another Sidewinder.[7]

At China Lake, little by little, active research and development activities were replaced by those *monitoring* research and development taking place elsewhere. Each person interviewed had his own favored symptoms of decline. For Fred Davis, it was the lack of technicians to implement ideas; for Ernie Cozzens, it was the amount of money spent on administrative expenses; for Hank Blazek, it was the small amount of discretion in spending that the average China Laker had, in comparison to the private sector. But it all added up to one thing: less hands-on technical work and more "paper engineering."

A Struggle over Principles

The changes in China Lake can be traced in parallel with China Lake's principles of operation. First formulated by L. T. E. Thompson and Cdr. Chick Hayward in 1946, the principles were meant to make clear that civilian scientists worked in *partnership* with the military at China Lake, not in subordination to it. The principles also made clear that top leadership was shared between a commander and a technical director, rather than having one report to the other. The driving force behind the principles was to create the kind of environment that would attract the very best in scientific and engineering talent.[8]

In 1955 the principles were reformulated, with only minor changes. They had no legal standing. China Lake was a military base and was subject to military law. More important than the exact words used to express them were what they stood for.

In 1974 Rear Adm. Rowland Freeman III made major changes to the principles, changing their spirit and substance to accord with his own views, which were that the military was to be firmly in charge of the civilians. In 1976 Freeman did away with the principles altogether. Under Freeman and his successor Rear Adm. William L. Harris, the military essentially created a line organization, breaking a tradition established thirty years before. Leroy Riggs described his reaction:

> You know, the day Freeman tore up the Principles of Operation, I thought Hugh Hunter would come unglued in the meeting. . . . Gil [Hollingsworth, the technical director] just said "Well, that sounds just fine to me. I'll sign it." If I had been Technical Director, I would have said, "Admiral, I've got a set of principles that were initiated by a set of great people in [1946], and we're going to Washington, you and I, and if necessary, I'll go as high as necessary." We still had a lot of people in Washington who knew what made this place great.[9]

In 1980 Technical Director Robert Hillyer resurrected the principles; they were rewritten and actually signed by the center commander but not reissued. Management considered the reissue too sensitive at that time. Hillyer's document was based on the principles in effect during the McLean years (1954–67). While Burrell Hays was technical director from 1982 to 1986, the center's Board of Directors went on a retreat in which many issues were discussed. One of the outcomes was a decision to issue an "operating philosophy" in 1984, along with the center's five-year plan. This operating philosophy was essentially the principles written up by Hillyer. In the summer of 1985, the rewritten principles were reissued as "operating principles." Hays, however, soon fell victim to politics and resigned under pressure.

These changes in the principles reflected a tug-of-war between the station's unique charter as a "civilian–military partnership" versus a more traditional role as a navy laboratory carrying out the navy's wishes. The differences cut to the heart of conflicts between McLean's philosophy of weapons development and the traditional one he fought.

The Changes Outside

What happened to China Lake was a microcosm of what was taking place in the navy as a whole. World War II shaped the birth of China Lake. The generation of officers and managers tested in that war had enormous self-confidence. The postwar navy R&D community was staffed by leaders like Adm. Deak Parsons and Capt. "Count" Ruckner who believed in success and in themselves. When Captain Ashworth took over at China Lake, he told McLean: "I'm here to support you."[10] To this generation, McLean was a symbol of their own success. It was such leaders, who consciously sought and cultivated excellence in the operations under their control, who created NOTS and protected it in the early years.

[259]

World War II passed, to be replaced by Korea and after some years, Vietnam. Korea and Vietnam provided a good deal of combat experience, but neither transformed the civilian R&D culture the way World War II had. Washington moved away from wartime culture and brought in more "rational" and bureaucratic forms of operation.

The change in orientation of the Washington bureau managers was matched by a change in skills. In the early days, McLean had China Lake and Washington support from high-ranking officers who recognized talent. Admiral Parsons approved Sidewinder because he and L. T. E. Thompson believed in McLean. In 1963 when Barney Towle as comptroller in charge of R&D for the Bureau of Weapons was asked for funds to support the two-man "Moray" fighter submarine at China Lake, he managed to scrape together the money:

And Admiral Ashworth, near the end of probably 1963, told me he needed $11 million, which now the Secretary of Defense couldn't sweep up in three weeks; but I could then, because of the rules, not because of my great stealing ability. (At the end of the fiscal year we always had a sweep-up of monies. At that time we could juggle money around much more freely.) He needed them, and he needed them for Bill McLean, because he was building a submarine. Moray!

. . . There was a hell of a lot of scream about a submarine being built in the desert.

Bill had the wonderful faculty of support from guys like P. D. Stroop and Ashworth and Ruckner. They had so much respect and confidence in him that it overcame their parochial feelings—which they didn't have in unusual measure; they had it in small amounts, I guess. Compared with most senior people they were *not* strongly parochial. The fact that it was a submarine and it was in the desert didn't mean a damn thing to them, because they thought that it was useful and workable, and they had confidence in Bill. So that was very impressive to me, that these guys—and I saw it in more than one senior officer—had that much confidence, and it transcended the normal barriers of appropriation.[11]

In 1967, when McLean left China Lake for San Diego, this generation of officers was being replaced by those with less experience and more book knowledge. By the 1970s the number of R&D personnel who had had combat or wartime experience was smaller. In Washington the new generation of managers was shaped more by the precepts of the Harvard Business School than by memories of Midway or Los Alamos. Capt. Charles Bishop, former commanding officer at the Naval Ocean Systems Center, described this change:

In the 1960s . . . there was still a good strong cadre of qualified people in the Navy . . . Bureau of Ships, Bureau of Weapons, Bureau of Aeronautics. In the 1980s I don't think it exists to that extent. What's happened is that in the ensuing twenty years talent that came in in World War II, young guys off the bench who'd developed the capabilities to manage and do things, retired. They have been replaced by a lot of younger people that don't have that background; the technical competence isn't there. So what you find out, now, if you go to an office in Washington, if there's a technical question, pick up the phone, and here comes a beltway bandit in. The poor guys in the bureaus, their time is spent on budgetary problems.[12]

Technically competent people remained in Washington, but they were a minority. Many of them remembered the old bureau system in the navy: although parochial, it nonetheless had many officials who knew what they were doing. Although the new systems commands often were better organized to cope with the complex high-technology problems of the twentieth century, the people who ran the systems commands seldom had the hands-on knowledge of the old bureau chiefs.[13]

The new managers, however, did have splendid means for micro-managing research at the laboratories. With bureaucrats who depended on contractors for real, hands-on expertise, the answer that worked was getting adherence to systems. Systems would allow those without real working knowledge to manage those who possessed such knowledge. At least, that was the theory.

Thus at the same time that Washington imposed ever more controls, those in the seat of power had steadily less understanding of the technological aspects of the operations they were watching over. China Lake did not need micromanagement; it needed intelligent direction and support. It had such direction and support through the 1960s. By the mid 1970s, however, such intelligent counsel was already getting scarce. Today it is very scarce indeed.

Changes in Engineering

Engineering itself was changing from the hands-on culture of World War II to a more analytical and scientific style. This style encouraged abstraction, computer-based simulation, and theory. Of course, the space program could not have been done without such modeling and simulation. But while a can-do culture requires hands-on expertise, an analytical one does not have this same "bias for action." Analysis was encouraged by the increasing sophistication of the vehicles themselves, becoming ever more complex and more expensive. Emphasis was placed on getting each weapon system to do more, at a higher price tag. Weapons stayed in development longer before they were field-tested, and the tests got more expensive.[14]

Each new generation of Sidewinders got fewer firings. Many questions were now answered through computer simulations. Table 19.1 shows the dramatic changes in missiles used for several Sidewinder versions.

Equally serious was the belief that weapons development had now been made "rational." According to Howie Wilcox: "There was basi-

Table 19.1. R&D Firings of Sidewinder

	Research	Technical Evaluation	Operational Test and Evaluation	Total
AIM-9D	49	32	48	129
AIM-9L	21	20	28	69
AIM-9M	10	8	17	35

Source: "Naval Weapons Center: China Lake Readies Data for 1990s Sidewinder Upgrade," *Aviation Week and Space Technology,* 20 January 1986, 81.

cally an avalanche, which started about the end of the Sidewinder program, of impressions out of Washington, that you could calculate everything, that you could *know before you did it* whether it was going to be successful or not. And you only did, then, successful things. And this led to giant computer programs, and for a while, if it came out of a computer, hey, that was it!"[15]

These changes in engineering and management style meant that technical expertise shifted from Washington to private industry. If engineering was more rational, then it could be managed remotely. Increasingly, the Washington weapons manager was a person who made decisions about weapons systems that were developed by private firms. The role of government laboratories, then, became that of watchdog for Washington, "managing" the programs carried on by industry. But how could the laboratories be good managers of technology without hands-on experience in design? And how could Washington bureaucrats know whether they *had* good managers unless they knew what good management was?

Government Laboratories: Developers or Support Staff?

If some in the navy were bothered by China Lake's lack of orderliness and system, others believed that China Lake should not develop weapons at all—development was for private contractors. In aviation R&D virtually all systems were developed by large companies like McDonnell, Douglas, General Dynamics, and Lockheed. Strategic missiles also typically were developed by vendors. Ordnance was a more ambiguous area. Military laboratories might well research fundamental technologies, but private firms would then employ the technologies to develop the weapons.

Government laboratories, industry argued, were not for invention; they were for activities that supported invention and development in private industry. Whereas the army and the navy in the past had developed much of their own technology, the air force had decided that its new technology development would take place largely in industry, and air force laboratories would stimulate and monitor such research.[16] While weapons concepts might be evolved by air force laboratories, what McLean called "the building of ideas" should take place in industry. According to this view, government would only think up the problems. Private industry would create the solutions.[17]

So what should the labs do? Well, they could support industry through basic research. They could even, as China Lake showed through Burrell Hays's efforts, act as "systems managers," although industry wanted that role reduced, too. Washington bureaucrats were no more enthusiastic than industrial leaders about an aggressive China Lake. But to be a good "weapons system manager" China Lake needed personnel who could develop original designs. Without true design knowledge, their management of others' systems would be spineless and toothless. The station's actions on Sparrow, Condor, and similar systems showed that hands-on experience was vital even in the station's role of systems manager.

This was one of the major points on which Admiral Freeman differed from previous China Lake commanders—all of whom had come from the ranks of the Bureau of Ordnance, which was used to developing its weapons at government arsenals. Freeman, however, had come from the Bureau of Aeronautics, and BuAer was used to letting contractors develop its weapons. As far as Freeman was concerned, therefore, the Naval Weapons Center ought to get out of the business of development and into the business of systems management.[18]

Many in industry and government believed that rational management would bring about orderly acquisition of new technology with the aid of industry. To many, the Harvard Business School was the epitome of this rational approach. When Freeman said he wanted "cost, schedule, and performance in that order," he expressed this view. Yet many people doubted that an organization that could not develop a system would be good at managing it.[19]

While China Lake turned out to be an excellent policeman in regard to weapons system quality control, this role had contradictory aspects. China Lake was supposed to act impartially, but how impartial could it be if it were developing its own systems? Many considered the role of developer inconsistent with that of weapons system monitor. Obviously, too, many private firms had financial reasons for wanting to remove China Lake from competition. In its role as policeman, China Lake was encouraging second sourcing, a real menace to private profits, and the station's expertise was made even more dangerous when it developed systems that competed directly with those from industry.[20]

The Smart Buyer

The government now expects industry to develop and produce its weapons. Government, then, has become a "smart buyer." To do this, however, it must know the value of those weapons, and as Jack Crawford so aptly put it, the only way such knowledge can be gained is by a lot of hands-on experience. This case history suggests that a government that cannot develop its own weapons will not be good at buying them from others.[21]

The United States would not have developed the air-to-air missile it required without China Lake. This is most obvious in the development of Sidewinder, but it is equally true of the improvements in Sparrow III and many other missiles. China Lake was able to provide the proper benchmarks because its technical personnel could and did design weapons themselves. The hands-on expertise allowed China Lake to act as a wise umpire. Frank Knemeyer, former head of China Lake's Weapons Department, put it this way:

> You've got a government-industrial team. There's no question: industry produces *all* the hardware that the military uses. But somebody has got to decide the need and the concept and so on. . . . I say the government has got to be sure that there is an adequate technology base for all this. They should have demonstrated in theory and application all the technology. Government has to be the one that accepts

the thing, because if you build something, and I accept it and it doesn't work, it's my fault first, because I let you get by with something. There are too many things the government accepts that don't work. And the government has got to be sure that it's managed. OK, now who's going to do that on the government side?[22]

Who indeed? If government laboratories are going to act as judges, they [265] must be able, in principle, to duplicate the technical achievements of private contractors. And this can lead to competition. But how can a competitor be unbiased?

The truth is that an unbiased judge is not a realistic or attainable goal. Bias is a fact of life. Government laboratory personnel are going to have biases, just as anyone else will have biases. It is true that government jobs depend less on getting a system into production. But other factors can be strong. Organizational pride, technical orientation, and intellectual commitments are each likely to provide bias. Yet in the end these forms of bias are less dangerous than ignorance. Ignorance, too, is a bias, as is often shown by military officers whose careers have become dependent on noncombat "success" of a contractor's weapon system.

To be a "smart buyer," then you have to be smart, and that judgment is formed by hands-on engagement. Jack Crawford, now a senior weapons designer, shared this view:

> "Smart buyer" is almost a red flag for me. I'm sure the concept of the "smart buyer" comes from an analogy with consumer purchases. You do not need to know how to build an automobile or a TV set to be an intelligent purchaser of one. In that case, you need to know what performance you require and what is available in the marketplace. The problem in weapon development is that we are not selecting the best choice among an assortment of already produced items. The "smart buyer" of weapon development must have an in-depth understanding of the current state of the art . . . to make a knowledgeable choice among proposals. *That in-depth understanding comes only with hands-on experience with the technology.* "Smart buyer" is fine provided it's a by-product of your own technical competence at doing the job, so you're out buying the things you decided you need

to buy and doing the things that you've decided you can do better. But if you're simply training to be a buyer, competence will decline rapidly. Nobody who is really technically competent is interested in being a paper-shuffling reviewer of someone else's work. You will very quickly evolve into a bureaucracy that's just as bad as many of the Washington offices are now.[23]

Technical expertise is judgment based on experience. A far worse bias will exist when the government must decide, *with no ability to check,* which system to buy based on industry's presentation of the facts. It would be easy to cite cases in which misleading claims were accepted and promoted by government officials, but let us stick to two in the air-to-air field.[24]

First, the Falcon missile was subjected to constant competition from Sidewinder after the latter was developed. This competition was so intense that at one point Hughes was forced to close its Tucson plant for months and redesign the missile. Without this pressure, Falcon would have been even worse. Obviously, without such competitive shoot-offs Falcon would have appeared different to the government that bought it.[25]

Second, the Sparrow III missile was tested extensively at Point Mugu before it went into action in Vietnam, but Raytheon—the contractor—possessed instrumentation far superior to that of the government and strongly influenced the terms and the interpretation of testing. Firings generally put Raytheon's carefully nursed and hand-built prototypes in the middle of the performance envelope, so problems often went unnoticed. The net result was a fragile missile, because government tests never forced the contractor to make needed changes. The situation continued into the Vietnam War, until the Ault investigation laid it bare. Only then was Raytheon forced to improve its manufacturing.[26]

Truth emerges from the struggle between competing claims. Where weapons systems are concerned, this struggle is waged through tests. The shoot-offs, fly-offs, and sail-offs are the prelude to tougher operational tests in the field. While combat experience is the final test, it often comes too late to help the weapon's designer or user. This means that tests are all-important. The best guarantee of the honesty of weapons testing is the personal knowledge and experience of the

judges. *There is no substitute for the actual practice of engineering.*
Those without the tough lessons that experience teaches will not be
able to sort out good and bad.

Recently a damning indictment of academic professors of engineering who had never done any designing was published in the *Bent,* journal of the Tau Beta Pi engineering honor society. The article's author
pointed out many mistakes not only in lectures but also in engineering texts. These errors reflected the professors' lack of real-life design
experience. In one case, the author of a book used in more than one
hundred universities "didn't understand the SI system or hydrostatic
design. . . . The author never saw a [hydrostatic] vibrator work and had
no feel for real devices." The role of China Lake as intelligent judge of
weapons plans and tests, then, rested on its collective experience in
design, development, and testing. Such expertise did not come without hubris.[27]

Tom Amlie returned from one trip to Hughes while both NOTS and
Hughes were working on radar missiles:

> When I went to Hughes I was prepared to feel quite superior because
> I thought SARAH was superior to the [Falcon]-1D with which I was
> familiar and at which we had all been laughing. I wasn't laughing
> when I left after having seen the -3A. My impression is that these
> guys are good and they are serious about building guided missiles.
> They are real professionals whereas we are still a bunch of enthusi-
> astic amateurs. We can learn a lot from them.[28]

But the "real professionals" did not design the better missile. This
example shows the dangers of peer review when no hardware has
been produced. Which panel of experts would have chosen the China
Lake team over the Hughes team at the beginning? Sidewinder was
considered to be inadequate, even if it worked, and many experts were
agreed that it never would work. Yet the real, gritty, demanding, and
frustrating job of producing a working missile went different in real
life than it did on paper. Sidewinder became the world standard; Falcon failed.

Some peer review of Sidewinder was favorable. Many of its supporters, such as L. T. E. Thompson and Deak Parsons, believed in
Sidewinder because they believed in McLean—and they believed in

McLean because they recognized that he knew the realities of weapons development. It was seasoning through the experience of actual development, and in Parsons's case combat as well, that allowed them to be fair judges of the project and of McLean himself.

The really important thing was that China Lake existed at all. Without China Lake, there would have been no competitor to Sparrow III and Falcon. No doubt we would still have won the battle of the skies over Vietnam, but we would have lost far more American pilots than we did. Without Burrell Hays's expertise, the Ault report never would have allowed the dramatic improvements in design and manufacture that took place. China Lake permitted the government to act as an educated judge of its vendors' hardware and, when others' efforts were inadequate, to design its own.

Was China Lake Really Critical?

China Lake's skill was not confined to air-to-air missiles; the station designed Walleye and Shrike, both critical air-to ground weapons, many other forms of air-launched ordnance, and many, many other things. One could argue that alternatives to Sidewinder or Walleye or Shrike would have been developed elsewhere. Surely there were alternatives to Walleye. After all, Walleyes were not the only "smart bombs" dropped in Vietnam. The Bullpup air-to-ground missile— which Walleye largely replaced—had been developed by Martin Marietta, an aerospace company, and was the product of the standard "government requests, industry responds" system. Yet Walleye—another system developed without an operational requirement—was two orders of magnitude more accurate than Bullpup, and many of the other alternative smart bombs had been pioneered originally by a second government laboratory, the Naval Ordnance Laboratory Corona, whose roots went back to the National Bureau of Standards in World War II.[29]

It might also be argued that private industry is just as capable of producing a highly creative atmosphere as was China Lake. No argument there. The Lockheed Skunk Works provides an environment often acknowledged to be as good as anything in the world. But Lockheed's gem is a special case; a skunk works at a contractor usually puts the corporation's interest ahead of the national interest.[30]

The air force, unlike the navy, does not have laboratories that develop weapons systems. True, many aircraft concepts may have originated in an air force laboratory, but prototyping and development of aircraft and weapons are done by private firms. The air force laboratories test and evaluate, but development is largely left in the hands of the aerospace companies. Has the air force suffered from the lack of a China Lake? We just don't know. Not having done the experiment, the air force is in no position to judge what such development-oriented laboratories might yield.

[269]

Finally, critiques of weapons systems can be generated by a combination of whistle-blowers, government reformers, and an active press corps of investigative journalists. Sometimes such groups have been effective. The problem is that these groups can seldom generate technical alternatives or changes. They certainly can champion such alternatives when they arise, but only in rare instances can they exert leverage in getting completely different systems. The activities of the "tactical fighter mafia," a group of internal reformers at the Pentagon who shaped the design of the A-10, F-15, and F-16 aircraft, are an exception here.[31]

Such heroic activities usually fail, however, and even when they don't, they typically extract high costs from their participants. Furthermore, as development proceeds, the original insights may be lost as other forces are brought to bear. Such ad hoc groups are no substitute for the systematic oversight and pressure that a competent and independent-minded government laboratory can bring to bear. And of course a working prototype has an ability to convince that no verbal argument can match.[32]

Systems versus Functions

Associated with the developer versus manager issue was the project versus functional department struggle. Was China Lake's role to develop systems or to help others develop systems? McLean, Wilcox, and their associates were project people; they liked to develop whole systems. They tended to look down on the cats and dogs aspects of the station's role. While others tinkered and experimented to improve existing systems, the project group took pride in new systems. To their promoters, the weapons systems engaged the highest forms of creativity, fostered entrepreneurship, and provided visible proof that China Lake was superior. There is no question that this had an important impact on morale and not just for the project teams themselves. For many years, a huge banner in the Michelson machine shop proclaimed WE BUILT SIDEWINDER!

Yet much of the station's work was to provide technical support to industry in designing, testing, and evaluating industry products. The continuous improvements provided by the functional departments were very important. China Lake carried out fundamental research that would have been expensive for industry. And once China Lake found something out, the knowledge could be shared with any element of private industry. As a government laboratory, sharing was China Lake's job. While these efforts were less visible to outsiders, those involved in them were very proud that China Lake represented a center of technical expertise.

The balance point for China Lake shifted over time, with changing technical directors and station commanders, and the various pressures exerted by external forces. In the mid-1970s the balance point shifted away from projects and more toward China Lake's other functions. In time "smart buyer," "center of expertise," and "contract monitor" were exactly what the Naval Weapons Center became. Projects still continued to come out of the laboratory, as well as the cats and dogs applications of the technical expertise that many saw as a national treasure. But there were no more Sidewinders or Shrikes.

A Contrast with Point Mugu

In 1992 China Lake and Point Mugu merged to become the Weapons Division of a new Naval Air Warfare Center. China Lake lost its technical director, and the admiral running the center was shifted to Point Mugu. Air Test and Evaluation Squadron VX-5 at China Lake and VX-4 at Point Mugu also merged to become VX-9. As the labs merged, a very curious thing happened.

To understand what took place, one must consider the two cultures. Even stripped of many design functions, China Lake was still proud and self-confident. Civilians at China Lake were used to doing things their own way and getting away with as much as they could. Civilians at Point Mugu, however, had always been subordinate to the military;

those less willing to be subordinate generally left. Thus China Lake bred civilians who were notably more aggressive and were more used to making their own decisions.

Just after the merger, appointments to run departments in the merged laboratory structure were supposed to be split roughly fifty-fifty between China Lakers and the Mugu-ites. Gradually, however, the departments got taken over by department heads from China Lake.[33]

[271]

This juxtaposition of the two laboratory cultures illustrates an important fact of life: Point Mugu's culture was that of a laboratory that did not build its own weapons. While both organizations began as testing organizations, China Lake was strongly shaped by its development capability and its success in building Sidewinder, Walleye, Shrike, and many similar systems. If China Lake tested products that it thought were inadequate, it could do more than complain; it could redesign them. In extremis it might in fact offer to build alternative systems that were better—as it did with Sidewinder. Furthermore, China Lake could and did develop second sourcing if it thought the quality was too low or the price was too high. In other words, China Lake reacted aggressively to fix problems.

Point Mugu operated differently. While technical integrity was just as great at Point Mugu as at China Lake, Mugu's resources for action were fewer. When Point Mugu engineers refused to approve Sparrow I's release to the fleet, the commanding officer changed the memo to read that the system *was* approved for release. Whatever protest took place was private.[34] At China Lake similar attempts to circumvent technical honesty were strongly opposed by technical directors. Two TDs—Tom Amlie and Burrell Hays—were sacked when they refused to give in. More often, however, China Lake civilians were successful in getting their way. Thus even though Sparrow III had been developed through tests at Point Mugu, it was China Lake that eventually reshaped it.

China Lake believed in itself because of what it had done, and its vision of technical integrity was accepted by many higher officers in the navy. China Lake graduated a whole series of senior officers who became influential in the navy weapons bureaucracy. Among them were Admirals Ashworth, Hayward, Moran, Stroop, and Levering Smith. These officers, and those who shared a vision with them (notably John Rexroth of NavAir), protected and promoted China

Lake's activist style. When the protective cover became thin, so did China Lake's influence.

The ability to act was thus connected to the ability to think. When an organization is able to act on its perceptions, this is a strong motivation for critical thought. When an organization, for whatever reason, cannot act on its perceptions, it loses the will to think. One wonders how much this has happened to China Lake.

In Perspective

China Lake was created out of desperate need by far-seeing people aware of both its immediate value and its future promise. It was developed into a highly versatile laboratory good at virtually every aspect of research and development related to weapons. It perfected the use of test equipment. It trained and developed outstanding researchers and engineers. It acted as a "community of good judgment" in regard to processes and products. It developed an internal sense of fitness and appropriateness and measured everything against that standard. It brought tough judges—the users—in to evaluate its products and listened carefully to what they had to say. It had the great good fortune to have outstanding leaders and intelligent direction from above. For many, it was Camelot.

Out of this laboratory poured the designs for weapons that assisted the navy in air warfare: guided missiles, smart bombs, warheads, explosive devices, guidance units, launchers, software, even parachutes. China Lake did many things that it was asked to do, but it did other things that it was not asked to do, and even some things it was ordered not to do. Sidewinder was one such thing. It worked on assigned problems, but if it did not like the assignment given, it produced something else. When it wanted to launch a space satellite, it did so. Its personnel felt they could do *anything*. Again and again, they proved able to deliver on that promise.

They were not perfect, but they drew pictures on paper, cut metal, hustled things out to the test stand, and fired them off. They tinkered, tested, judged, and perfected. If others complained about their standards, they tried to educate or outfox the complainers. They tried to do things they shouldn't have tried. Things blew up at the wrong time; planes crashed. People died. They buried their dead and went on.

Not everyone fit in; people left. The ones that stayed built a community, thinking, living, breathing, romancing, marrying and divorcing, raising children. They joined clubs, built things in their garages, and went up into the mountains to hike, ski, or swim. But most of all they talked. They talked at work; they talked after work, over the back fence, at the Officers Club, at cocktail parties. Sooner or later the talking led to ideas for things. They returned to the office or the laboratory and figured out how to bring their ideas into fruition.

China Lake was a military base and a civilian laboratory. The partnership of naval personnel and civilian tinkerers created a unique environment. It was not a stable system; it could not afford to be. To manage this unstable system required management of high caliber at all levels, and mostly China Lake got that kind of management. There were conflicts. Some viewpoints prospered and others were shunted aside.

It was fragile. Big, sprawling, creative, organic, and persistently independent-thinking, it was still vulnerable. The navy is a bureaucracy, and it wanted China Lake to be a bureaucracy, too. The more success China Lake had in its activities, the more powerful the pressures were to bring it into line. Every highly creative place exists on borrowed time. Bill McLean could protect China Lake when he was technical director. When he left, protecting it was harder.

Outside forces became stronger. In the end China Lake was forced into line. Rules and regulations proliferated. China Lake lost its technical director and was merged with Point Mugu. The laboratories are still filled with capable people, but their freedom of action is reduced. Those who want to disobey are forced to lie low. Today the lab is still a beehive of capabilities, but its continued existence is in doubt. At a time when bases are closing and military budgets are being reduced, it is an open question whether China Lake will survive. Let us have the wisdom to maintain this organization that has served us so well, and the courage to create others like it in the future.

Notes

Preface

1. Arthur Squires, *The Tender Ship: Government Management of Technology* (Boston: Birkhauser, 1986), 13–46.

Chapter 1. The Weapon Nobody Asked For

1. See Gregory Vistica, *Fall from Grace: The Men Who Sank the U.S. Navy* (New York: Simon and Schuster, 1995), 106–7.
2. Lt. Larry Muczynski, "Pilot's Account," in Bert Kinzey, *F-14A and B Tomcat* (Blue Ridge Summit, Pa.: Tab Books, 1982), 42–47.
3. DeVirl A. Kunz, quoted in "Bud Kunz: A Center Pioneer Retires," *N.O.S.C. Outlook,* 7 September 1979, 1. See also Robert Gannor, *Hellions of the Deep: The Development of Torpedoes in World War II* (University Park: Pennsylvania State University Press, 1996), 174–75.
4. Gordon W. Prange, Donald M. Goldstein, and Katherine V. Dillon, *Miracle at Midway* (New York: McGraw-Hill, 1982), 176, 308–12, 388.
5. Jeffrey Ethell and Alfred Price, *Air War South Atlantic* (London: Sidgwick and Jackson, 1984), 253–54.
6. See Bruno Latour, *Science in Action: How to Follow Scientists and Engineers through Society* (Cambridge, Mass.: Harvard University Press, 1987).

Chapter 2. The Gadgeteer

1. This awkward quality is evident in the few glimpses afforded of McLean in motion in the video *Secret City,* a Naval Weapons Center Museum documentary on the history of China Lake.
2. Howard Wilcox, comments on draft.
3. William B. McLean, "Autobiography," McLean Collection, David Taylor Laboratory. This is a brief document written for his induction into the National Academy of Sciences, about June 1974.
4. McLean, "Autobiography," 1, 2.

5. McLean's youthful curiosity also stayed with him in later life. Marooned in Washington over a weekend with colleague Thomas Amlie, he and Amlie decided to pass the time at what is now the Smithsonian's National Museum of American History. McLean noticed a linotype machine and began to study it; Amlie wandered off. When he returned an hour later, McLean was still at the linotype, every part of which he now understood— and which he proceeded to explain to Amlie in detail. "I am convinced," Amlie told me, "that he would have spent days looking at it if he had to in order to understand it" (Thomas S. Amlie, letter to the author, 1986).

6. "From Gadgeteering, A Guided Missile Weapon with Complex Functions but Simple in Design," *Rocketeer,* 19 October 1956, p. 1.

7. All three boys succeeded: William as a scientist, John as the president of Continental Oil, and Robert as a pastor.

8. Telephone interview with George Jennings, 30 July 1991. Unless otherwise indicated, all interviews were conducted by the author.

9. Telephone interview with Roderick McClung, 1 April 1989, and interview 22 February 1995.

10. Telephone interview with O. H. Shoemaker, 2 June 1990.

11. McLean, "Autobiography," 3.

12. Judith R. Goodstein, *Millikan's School: A History of the California Institute of Technology* (New York: W. W. Norton, 1991), 94.

13. McLean, "Autobiography," 3; comments by William A. Fowler, 27 August 1991, on a draft of this chapter. Fowler was himself a graduate student of Lauritsen's and directed rocket research at China Lake during World War II. He received the Nobel Prize for Physics in 1983.

14. Interview with Claire and Robert Blohm, 3 March 1987. The word "gadgeteering" is used extensively in a major article announcing the Sidewinder in China Lake's publication the *Rocketeer,* 19 October 1956, "From Gadgeteering, A Guided Missile Weapon with Complex Functions but Simple in Design." According to Howard Wilcox, McLean often used the term "gadgeteer" to describe himself.

15. Interview with Haskell Wilson, 3 March 1987; interview with George Jennings, 30 July 1991.

16. Interview with LaVerne McLean, 14 January 1988.

17. Telephone interview with Henry Swift, 4 May 1991.

18. Telephone interview with Henry Swift, 4 May 1991.

19. See Irwin Stewart, *Organizing Science for War* (Boston: Little Brown, 1948), 86.

20. Telephone interview with Henry Swift, 4 May 1991; telephone interview with Jacob Rabinow, 19 November 1995.

21. Interview with Jacob Rabinow, 1 March 1988. Unless otherwise indicated, all quotations by Rabinow in this chapter are from this interview.

22. McLean later acknowledged the inventor's abilities in recommending Jacob Rabinow for the Harry Diamond Award of the IEEE (McLean to Karl Willenbrock, 17 February 1976).

23. Stanley Atchison, former technical director of the Naval Ordnance Laboratory at Corona, also worked in the wartime bureau. Atchison arrived at the Bureau of Standards in November 1942. In a 5 January 1993 letter to the author, he said it was just as informal as Rabinow remembers.

[277]

24. The Lockheed "Skunk Works" under Kelly Johnson created the P-80 Shooting Star jet fighter, the U-2 and SR-71 spy planes, the F-117 stealth fighter, and many other pieces of effective materiel. Design teams for the P-80 and other projects were kept lean but the group's informal and intuitive operation, as well as Johnson's brilliant leadership, allowed creativity to soar. See Clarence L. "Kelly" Johnson with Maggie Smith, *Kelly: More Than My Share of It All* (Washington, D.C.: Smithsonian Institution Press, 1985).

25. Jacob Rabinow. "The Individual in Government Research and Innovation," in *Innovation and U.S. Research: Problems and Recommendations,* ed. W. Novis Smith and Charles F. Larson (Washington, D.C.: American Chemical Society, 1980), 161.

26. Interview with Jacob Rabinow, 1 March 1988.

27. Some of McLean's colleagues did not find his bad ideas so easily discarded. Consider McLean's interactions with Jagiello over aerodynamics and launchers in chapter 7.

28. Comment by Franklin Knemeyer and Leroy Riggs on manuscript.

29. This anecdote was remembered in various forms by his family and friends. Throwing up his hands and announcing "That's it!" was very common when he had solved a problem, according to his son, Mark McLean, in an interview with the author 24 July 1988. A similar anecdote is recorded in an Oral History interview with Adm. Thomas H. Moorer, by Elizabeth Babcock, S-208, 4 May 1992.

30. William McLean, excerpt from acceptance speech for American Ordnance Association Blandy Medal, 24 May 1960.

Chapter 3. The Problem Takes Shape

1. James Phinney Baxter III, *Scientists against Time* (Cambridge, Mass.: MIT Press, 1968); Irwin Stewart, *Organizing Science for War* (Boston: Little, Brown, 1948); Daniel J. Kevles, *The Physicists: The History of a Scientific Community in Modern America* (Cambridge, Mass.: Harvard University Press, 1987), 302–23.

2. Vannevar Bush, *Pieces of the Action* (New York: William Morrow, 1970); the quotation is on 31–32. See also G. Pascal Zachary, *Endless Frontier:*

Vannevar Bush, Engineer of the American Century (New York: Free Press, 1997).

3. The atomic bomb was developed by the Manhattan Project, in which Bush was involved, but OSRD participated only in the initial stages. See Stewart, *Organizing Science,* 120–21.

4. Ralph Baldwin, *The Deadly Fuze: The Secret Weapon of World War II* (San Rafael, Calif.: Presidio Press, 1980), 105.

5. Jay Miller, *Lockheed's Skunk Works: The Official History,* 2d ed. (Leicester, U.K.: Midland Publishing, 1995). See E. T. Wooldridge Jr., *The P-80 Shooting Star: Evolution of a Jet Fighter* (Washington, D.C.: Smithsonian Institution Press, 1979). Johnson originally hated the term "Skunk Works" and reconciled himself to it only after the passage of time.

6. Clarence L. "Kelly" Johnson with Maggie Smith, *Kelly: More Than My Share of It All* (Washington, D.C.: Smithsonian Institution Press, 1985); Amrom Katz, "Kelly Johnson on the Weapons Acquisition Process," Rand Document 20444-PR, 30 June 1970. Permission to cite this document was given to the author by Amrom Katz. Thomas Peters and Robert H. Waterman, *In Search of Excellence: Lessons from America's Best-Run Companies* (New York: Warner, 1982). The Peters and Waterman book made the term "skunk works" popular, and Peters holds annual "skunk camps."

7. Albert B. Christman, *Sailors, Scientists, and Rockets,* vol. 1 of *History of the Naval Weapons Center, China Lake, California* (Washington, D.C.: Government Printing Office, 1971), 73. See also Christman's *Target Hiroshima: Deak Parsons and the Creation of the Atomic Bomb* (Annapolis, Md.: Naval Institute Press, 1998).

8. Bush felt, however, that during the war the navy made great strides in learning to cooperate successfully with civilians in R&D. See Vincent Davis, *The Admirals Lobby* (Chapel Hill: University of North Carolina Press, 1967), 174–75.

9. Judith R. Goodstein, *Millikan's School: A History of the California Institute of Technology* (New York: W. W. Norton, 1991), 252.

10. Christman, *Sailors, Scientists, and Rockets,* 163–204.

11. Conway Snyder, "Caltech's Other Rocket Program: Personal Recollections," *Engineering and Science* 54, no. 3 (spring 1991): 2–13.

12. Jacob Rabinow comments in *Inventing for Fun and Profit* (San Francisco: San Francisco Press, 1990) that he had largely negative experiences at army arsenals. "They test; they follow tradition; they're conservative; they're safe—but originality is not one of their strong points," 63. This was particularly true of the Bureau of Aeronautics, where aircraft had typically been designed and developed by contractors. The Bureau of

Ordnance, by contrast, typically had developed its hardware in-house prior to World War II. On this point, see Taylor Peck, *Round-Shot to Rockets: A History of the Washington Navy Yard and the U.S. Naval Gun Factory* (Annapolis, Md.: U.S. Naval Institute, 1949); Berend Derk Bruins, "U.S. Naval Bombardment Missiles, 1940–1958," Ph.D. diss., Columbia University, 1981, 257.

13. J. D. Gerrard-Gough and Albert B. Christman, *The Grand Experiment at Inyokern,* vol. 2 of *History of the Naval Weapons Center, China Lake, California* (Washington, D.C.: Government Printing Office, 1978), 243–73. The third volume of the NOTS/NWC history, being written by Elizabeth Babcock, will be entitled *Those Magnificent Mavericks.* NOTS principles as printed in Gerrard-Gough and Christman, *The Grand Experiment at Inyokern,* 403.

[279]

14. Text of BuOrd Order No. 28-51, 22 June 1951, 1.

15. Naval historian Albert B. Christman notes that the military-civilian R&D partnership was not new. Such a partnership functioned at the Naval Experimental Station at New London in World War I and later was typical of the Naval Research Laboratory (A. B. Christman, letter to the author, 1 May 1989). Vannevar Bush worked at the Naval Experimental Station in World War I, and this may have shaped his OSRD laboratory policies (Kevles, *The Physicists,* 294). In his own book *Pieces of the Action,* Bush several times strongly emphasizes the necessity of military-civilian partnerships in weapons research.

16. L. T. E. Thompson, "Remarks on the Organization and Operation of NOTS, 1945–1951," speech delivered before Senior Personnel Conference, NOTS, China Lake, California, 7 August 1951. McLean credited the success of the Walleye guided bomb program to NOTS's unique combination of thinking and testing facilities (McLean, Oral History interview with Albert Christman and R. G. Douglas, S-88, 16 November 1973, 12).

17. Baldwin, *The Deadly Fuze,* prefatory page.

18. Telephone interview with Emory L. Ellis, 14 December 1995, with following letter to the author.

19. Comments on draft by Elizabeth Babcock, August 1995.

20. Comments on draft by LaVerne McLean, August 1995. Interview with William McLean by Albert B. Christman, NWC Oral History 5313-S-97, July 1975, 2.

21. William McLean interview by Albert B. Christman, July 1975, 2.

22. Telephone interview with Simon Ramo, 28 July 1991. Dr. Ramo was responsible for the development of the Falcon missile at Hughes Aircraft until 1953.

23. Telephone interview with Simon Ramo, 28 July 1991.

24. Telephone Interview with Simon Ramo, 28 July 1991; Kenneth Schaf-
fel, *The Emerging Shield: The Air Force and the Evolution of Continental Air Defense, 1945–1960* (Washington, D.C.: Office of Air Force
History, 1991), 102.

25. Telephone interview with Fred Davis, 26 August 1991.

26. Unless otherwise indicated, all information in this chapter relating to
Colonel Scheller's activities comes from a telephone interview with him
on 27 March 1990.

A February 1961 Morton Thiokol fact sheet for GAR-3A radar Falcon
(declassified) indicates an 8.6-pound warhead with 5 pounds of explosive ("Characteristics Summary: GAR-3A"). A June 1963 fact sheet
quotes a weight for the infrared Falcon warhead ("including fuze") as
8.4 pounds, with only 2.75 pounds of explosive ("Standard Missile
Characteristics: GAR-2B Falcon"). Bill Gunston, *Modern Airborne Missiles* (New York: Prentice-Hall, 1986), 30, cites a warhead size of 29 to
40 pounds, but this is for later models, which were larger than the original models. In one "gunnery" test Falcons that came within twenty feet
were scored as a "hit," a very generous assumption indeed for a 5-pound
warhead. See Philip J. Klass, "Genies, Falcons Fired in Gunnery Meet,"
Aviation Week, 3 November 1958, 28–29.

27. Ray Wagner, *American Combat Planes,* 3d ed. (Garden City, N.Y.: Doubleday, 1982), 466–67.

28. See Constance M. Green, Harry C. Thomson, and Peter C. Roots, *The
U.S. Army in World War II,* vol. 1, *The Ordnance Department: Planning Munitions for War* (Washington, D.C.: Department of the Army,
1955), 437.

29. Interview with Jack Crawford, 25 August 1989; Henry Swift, letter to the
author, 15 June 1991.

30. "Inquiry Set on Failure of Jets to Down Drone," *New York Times,* 20
August 1956, 23.

31. Quoted in Mike Spick, *All-Weather Warriors: The Search for the Ultimate Fighter Aircraft* (London: Cassell, 1994), 115.

32. Telephone interview with Col. Donald Scheller, 3 July 1992.

33. Interview with William B. McLean by Albert Christman, NOTS Oral
History 5313-S-97, July 1975, 7.

34. Henry Swift, letter to the author, 15 June 1991.

35. William B. McLean. "The Sidewinder Missile Program," in *Science,
Technology, and Management,* ed. F. Kast and J. Rosenzweig (New York:
McGraw-Hill, 1963), 167–68.

36. McLean, "The Sidewinder Missile Program," 168.

37. William B. McLean testimony in *Weapons System Acquisition Process,*
Hearings before the Committee on Armed Services, U. S. Senate, 92d

Congress, December 1971 (Washington, D.C.: Government Printing Office, 1972), 230.

38. William B. McLean, "The Sidewinder Missile Program," 169–70; emphasis added. The thoughtful reader will notice here the kind of mental powers McLean took for granted.

39. Deborah Dougherty, "Interpretive Barriers to Successful Product Innovation in Large Firms," *Organization Science* 3, no. 2 (May 1992): 179–202.

40. Frederick A. Darwin, letter to the author, 30 April 1990; emphasis in the original.

41. Interview with William B. McLean by Albert Christman and R. G. Douglas, China Lake Oral History, 16 December 1973, 6.

42. The phrase "license to think" is borrowed from an article by William Shockley in which he quotes Enrico Fermi on the "will to think" (Shockley, "The Invention of the Transistor—An Example of Creative-Failure Methodology," in *The Public Need and the Role of the Inventor,* ed. F. Essers and J. Rabinow [Washington, D.C.: Government Printing Office, 1974]), 62.

[281]

Chapter 4. The Wrong Laboratory

1. Interview with Vice Adm. William Moran, 18 July 1988, 21.

2. Interview with Robert G. S. Sewell, 23 February 1993.

3. Interview with Newton Ward, 3 March 1987.

4. Telephone interview with Jacob Rabinow, 16 November 1994, 1.

5. Henry Swift, letter to the author, 15 June 1991, 3; Henry Swift, letter to author, 16 June 1991.

6. According to Howard Wilcox, this group would have included Sidney Crockett, Roger Estey, Bill Guy, Luke Biberman, Ted Whitney, Don Duckworth, Larry Nichols, Pauline Rolf, and others.

7. Interview with Gilbert Plain, 8 January 1988.

8. On 24 February 1949 a priority telex was sent from A. Vazsonyi to A. L. Bennet [sic] "GUIDED MISSILES PROGRAM CANCELLED. STOP ALL OF YOUR WORK ON NOTS AIR TO AIR MISSILE." A copy of this communication was given to the author by Cdr. Wade Cone.

9. Interview with Vice Adm. Levering Smith by Howard Wilcox, 24 April 1989.

10. Theodore Franklin Gautschi, *An Investigation of the Management of the Research and Development Process,* NOTS AdPub 112 (China Lake: U.S. Naval Ordnance Test Station, August 1962), 109.

11. Rear Adm. Parsons did not become deputy chief of the Bureau of Ordnance until March 1952. But in the immediate postwar period he belonged to three critical committees: the Military Liaison Committee of the Atomic

Energy Commission; the Weapon Systems Evaluation Group; and the Atomic Defense Division in the Office of the Chief of Naval Operations. This may be a partial explanation of Parson's influence even before his role as deputy chief. See Albert B. Christman, *Naval Institute Proceedings*, January 1992, 56–61. William B. McLean, oral history interview by Albert Christman, S-88, July 1975, 9.

12. Interview with Rear Adm. Thomas J. Christman, 25 March 1993, 3.

13. The conflict is described in Rear Adm. D. S. Fahrney's long typescript, "The History of Pilotless Aircraft and Guided Missiles," written about 1956 and located in the archives of the Naval Historical Center in Washington, D.C.

14. J. D. Gerrard-Gough and Albert B. Christman, *The Grand Experiment at Inyokern*, vol. 2 of *History of the Naval Weapons Center, China Lake, California* (Washington, D.C.: Naval History Division, 1978), 280.

15. *Days of Challenge, Years of Change: A Technical History of the Pacific Missile Test Center* (Washington, D.C.: Government Printing Office, 1989).

16. On 18 December 1950, Keller decided to "super-accelerate" the development and production of the three missiles, expecting production of a thousand apiece of each kind. See Berend Derk Bruins, "U.S. Naval Bombardment Missiles, 1940–1958," Ph.D. diss., Columbia University, 1981, 173–74.

17. David A. Anderson, "Sparrow Keeps Pace with Target and Aircraft," *Aviation Week and Space Technology*, 20 August 1962. Just to develop a factory for the Sparrow I, Sperry was given two grants totaling $25 million by the Bureau of Aeronautics. This would be a huge amount of money today. See Glenn Bugos, *Engineering the F-4 Phantom II: Parts into Systems* (Annapolis, Md.: Naval Institute Press, 1996), 78.

18. Max White, "The Battle of the Sparrows," *Launchings* (winter 1989). *Launchings* is the newsletter of the Missile Technology Historical Society.

19. According to Max White, while the lower echelons at Point Mugu recommended against releasing Sparrow I, management overruled them and prematurely released the missile for manufacture. Naturally, shipboard response to the missile was quite negative (interview with Max White, 17 June 1995).

20. Interview with Max White, 17 June 1995.

21. Telephone interview with Thomas Phillips, former chief of Sparrow III missile development and later chief executive officer of Raytheon, 15 March 1994.

22. Otto J. Scott, *The Creative Ordeal: The Story of Raytheon* (New York: Atheneum, 1974), 227. Actually, the interception was done by the Sky-

lark, a later version, according to Max White (interview, 17 June 1995). Fahrney, "History of Pilotless Aircraft and Guided Missiles," 572.

23. Scott, *Creative Ordeal,* 224, 231.

24. Scott, *Creative Ordeal,* 238, 253.

25. Interview with Bert Alton, 26 March 1995; interview with Max White, 17 June 1995.

26. Walter J. Boyne, *Phantom in Combat* (Washington, D.C.: Smithsonian Institution Press, 1985), 32. According to Max White, Sparrow test pilot Lt. C. C. Andrews had a major role in convincing McDonnell Aircraft that it should design a plane around the Sparrow III. According to White, most aircraft firms at this time took the attitude "We don't want your goddamn missiles; they're just a pain in the ass" (interview with Max White, 17 June 1995).

[283]

Chapter 5. Struggles with Infrared

1. Material in this section comes largely from an interview with Lawrence Nichols by David Rugg in about 1970, and two interviews with Nichols by the author, 2 August 1992 and 9 August 1992. Personal communication from Franklin Offner. Rear Adm. D. S. Fahrney, "History of Pilotless Aircraft and Guided Missiles," 1224, typescript, Naval Historical Center, Washington, D.C. Henry Swift remembers that a scientist from Eastman Kodak first started talking up infrared guidance to McLean. At that time Eastman Kodak was working on infrared fuzes (telephone interview with Henry Swift, 4 May 1991).

2. Interview with Newton Ward, 3 March 1987. In an interview with Albert Christman, in July 1975, McLean recalled Eastman Kodak's infrared system for bombing; I think this is a reference to the Dove. The negative view of Dove generally expressed by members of the Sidewinder team is not echoed in Rear Admiral Fahrney's "History of Pilotless Aircraft and Guided Missiles"; see 1218–24.

3. Ralph Baldwin, *The Deadly Fuze: Secret Weapon of World War II* (San Rafael, Calif.: Presidio Press, 1980), 300–301.

4. When David Rugg was writing his manuscript "The Sidewinder Missile" in the early 1970s, he noted that Nichols was still working in infrared and was a recognized national authority on the subject (13). A copy of this MS is in the China Lake Technical Information Department.

5. Donald T. Duckworth, "An Air-to-Air Target Seeker," NOTS 201, 11 March 1949.

6. Interview with Lawrence Nichols by David Rugg in about 1970, China Lake Oral History S-198, p. F-4.

7. Interview with David J. Simmons, 20 February 1995.

8. Interview with William Woodworth, 21 August 1992. While the basic ideas for the A head were McLean's invention, detailed design was actually carried out by Donald Friedman at Avion.

9. Interview with Lucien Biberman, 23 July 1991.

10. Thomas S. Amlie, letter to author, 11 September 1994.

11. There is some dispute over who really invented the checkerboard reticle. Howard Wilcox claimed he invented it, but declined to patent it. The patent (Navy Case 15,865) was filed in the name of Lucien Biberman and Roger Estey.

12. Material for this section is largely derived from a memorandum written by Robert B. Larsen of Taylor, Roberts, and Hinds, attorneys, "Brief on Behalf of William B. McLean, in the United States Patent and Trademark Office before the Board of Patent Interferences, Interference No. 92,172, *McLean vs. Osborne*," 20 May 1975. This document was filed in relation to an attempt by Aerojet-General to invalidate the McLean patent. Sidewinder had demonstrated outstanding performance in the Vietnam War; McLean was in bad health; and Aerojet-General may have thought the moment right to press its suit. General Tire and Rubber had earlier worked with NOTS on the NOTS air missile, and so GT&R's involvement with missile guidance was probably an ongoing effort.

13. Interview with Gilbert Plain, 8 January 1988.

14. Interview with Gilbert Plain, 8 January 1988.

15. Memorandum from T. H. Moorer, NOTS Experimental Officer, to Fred Brown, NOTS Technical Director, 8 October 1951, 3.

16. Interview with Howard A. Wilcox by Elizabeth Babcock and Mark Pahuta, NWC Oral History S-196, 22 October 1991, 77.

17. Charles P. Smith, comments on draft, March 1994.

18. Interview with Lawrence Nichols by David Rugg in about 1970.

19. Interview with Lawrence Nichols by David Rugg.

20. Interview with Lawrence Nichols by David Rugg.

21. The quotation is from an interview with Donald K. Moore, 20 August 1989. See William B. McLean, "Sidewinder Missile Program," in *Science, Technology, and Management,* ed. Fremont Kast and James Rosenzweig (New York: McGraw-Hill, 1963).

22. Charles P. Smith, comments on draft, March 1994.

23. Interview with Douglas Wilcox, 28 February 1987; interview with Walter LaBerge, 22 September 1990.

24. Interview with Donald K. Moore, 20 August 1989.

25. For some contemporary accounts of the problems with computer programs predicting the effects of weapons, see James G. Burton, *The Pentagon Wars: Reformers Challenge the Old Guard* (Annapolis, Md.: Naval Institute Press, 1993), esp. 128–30.

26. Interview with Howard Wilcox, 7 January 1988.

27. Charles P. Smith, Sidewinder Group Interview, 14 March 1980, by William F. Wright and Bud Gott, China Lake Oral History, S-112, p. 6.

28. Art Fry (3M corporate scientist), personal communication.

29. Originally McLean intended to use two gyroscopes, one for the seeker and one for stabilizing the missile. See McLean Notebook entry for 20 November 1948. During one of the earlier visits to Hughes Aircraft, however, McLean was persuaded by Hughes engineers that if the missile could be made stable, only one gyroscope would be needed (interview with Howard Wilcox, 1 March 1987). [285]

30. "A Heat Homing Rocket," Technical Memorandum 452-5, Aviation Ordnance and Test Department, U.S. Naval Ordnance Test Station, 20 June 1949.

31. See letter from L. T. E. Thompson [RexdL/TET, BuOrd] to Dr. W. R. Brode, NOTS, 15 August 1949. Dr. Brode was staying in Thompson's house at the time (Thompson papers at China Lake).

32. Gilbert J. Plain, letter to Howard Wilcox, 23 August 1981. According to Donald E. Carr pit vipers can detect a temperature difference of 1/1,000 of a degree centigrade (Carr, *The Forgotten Senses* [Garden City, N.Y.: Doubleday, 1972], 303).

33. Naval Speedletter to Capt. W. R. Vieweg, Commander NOTS, from Capt. M. R. Kelley, Bureau of Ordnance, 7 February 1951, serial number 16854; Joe Hughes, "Sidewinder Missile Story: Evading the Budget Knife," *San Diego Tribune,* 24 August 1981. The story is an interview with Howard Wilcox. For whatever reason, Wilcox did not mention this story to me directly.

34. The 1 March 1951 telephone directory for NOTS. Supposedly, an official telephone query to Cdr. Thomas Moorer, the experimental officer, prompted the numbers. "Well," the caller wanted to know, "what is the report number of your guided missile feasibility study?" Moorer and his assistant, Lt. Cdr. Bill Moran, thought fast—Moorer knew that such studies "couldn't be called just number one and number two." Moran immediately suggested they use the last three digits of their telephone extensions as the feasibility study numbers: they sounded official. See interview with Vice Adm. William J. Moran by Leroy Doig III, NOTS Oral History S-187, 5 December 1990, 31. This story, however, does not take into account the fact that Local Project 602 was apparently created a year before Feasibility Study 567.

Chapter 6. The Weapons Shop

1. Interview with Rear Adm. Thomas J. Christman, 25 March 1993, 3. Apparently, Admiral Parsons, then deputy director of the Bureau of Ordnance, decided on this piece of public relations (Elizabeth Babcock, comments on manuscript).

2. Memorandum from Head, Aviation Ordnance Department, to Technical Director, NOTS, 26 May 1950. Rear Adm. Thomas J. Christman's copy in possession of China Lake's Technical Information Department.

3. Memorandum from Head, Aviation Ordnance Department, to Technical Director, Naval Ordnance Test Station, 25 October 1950, 1. The intelligent fuze was actually for the HPAG (high-powered air-to-ground) rocket rather than the HPAA (high-powered air-to-air) one.

4. Jonathan Norton Leonard, "Birds of Mars," *Newsweek,* 21 May 1951, 82–90.

5. Howard Wilcox, "Reminiscences," part 1, 7–8. This four-part typescript was written especially for this book; hereafter cited as Wilcox, "Reminiscences," with part and page numbers.

6. Interview with Wade Cone, 19 July 1988.

7. Stanley Davis and Paul R. Lawrence, *Matrix* (Reading, Mass.: Addison-Wesley, 1977). Separating administrative and operational control creates similar problems with military units.

8. Robert Blaise, "The Theory and Practice of Creative Environments," manuscript, 25 August 1991, 9.

9. Evelyn Glatt, *The Demise of the Ballistics Division* (Berkeley: University of California Institute of Governmental Studies, November 1964), 14.

10. Ralph Katz and Thomas Allen, "How Project Performance Is Influenced by the Locus of Power in the R&D Matrix," in *The Human Side of Managing Technological Innovation: A Collection of Readings,* ed. Ralph Katz (New York: Oxford University Press, 1997).

11. General Momyer notes that most MiG kills in Korea and most fighter kills in World War II were achieved from the six o'clock position, that is, from behind (115). This would imply that an attacker with serious head-on capabilities would have a strong advantage (William W. Momyer, *Air Power in Three Wars,* privately published, 1978).

12. Charles P. Smith, Sidewinder Group interview by William F. Wright and Bud Gott, 14 March 1980, 20.

13. Interview with Robert Blaise, 25 March 1989, 2.

14. Burton Klein, letter to Eric Gustafson, 2 March 1962.

15. Telephone interview with Donald Friedman, 20 February 1994.

16. Telephone interview with Donald Friedman, 20 February 1994.

17. Howard A. Wilcox, comments on manuscript.

18. Doris M. Condit, *History of the Office of the Secretary of Defense,* vol. 2, *The Test of War* (Washington, D.C.: Historical Office of the Secretary of Defense, 1988), 474–76.

19. Keller was selected because the armed services had proven themselves unable to make "tough decisions" about which missile programs to cancel. Keller had been a "can do" industrialist in World War II and was expected

to get things moving. See Herbert F. York and G. Allen Greb, "Military Research and Development: A Postwar History," in *Science, Technology, and National Policy*, ed. Thomas J. Kuehn and Alan L. Porter (Ithaca: Cornell University Press, 1981). Wilcox, "Reminiscences," part 2, 9.

20. Interview with Lucien Biberman, 23 July 1991. Biberman remembers this episode happening as early as 1950. This was, however, some years before "Engine Charlie" Wilson became secretary of defense.

Chapter 7. Systems Engineering

1. Bernard Jaeger, letter to the author, 21 February 1994.
2. Henry Swift, letter to the author, 15 June 1991.
3. This was a point made by Walter LaBerge in a telephone conversation with the author on 22 September 1990.
4. Evelyn Glatt, *The Demise of the Ballistics Division* (Berkeley: University of California Institute of Governmental Studies, November 1964), 11.
5. Bernard Jaeger, letter to the author, 21 February 1994. Jaeger was present at the meeting between Wilcox and Jagiello.
6. Charles P. Smith, letter to the author, 13 June 1996.
7. For information on user resistance, see James Fallows, *National Defense* (New York: Random House, 1981).
8. W. F. Cartwright and Edwin G. Swann, "Guidance and Control," *Weapons* (China Lake, Calif.) 1, no. 1 (1959): 20.
9. Wilcox, "Reminiscences," part 1, 12.
10. Reference to NVD 1269, not personally examined by the author.
11. H. A. Wilcox and Lt. T. J. Christman, "Status Report of Sidewinder Program as of August 1, 1953," China Lake: Naval Ordnance Test Station, 30 October 1953, 22.
12. Interview with Fred Davis, 28 July 1991.
13. Thomas Amlie, letter to author. The washing machine story in relation to nutation problem is also mentioned in McLean interview, 16 November 1973, 4. There were many accounts as to whose washing machine it was. The version reported here is only one of them.
14. Interview with Leroy Riggs, 16 January 1988. McLean also commented to Riggs that "the best servo was the human brain." Once, when McLean was testing out an experimental parawing (a directional parachute) on Lake Isabella, aerodynamicist Dick Meeker nearly drowned trying to prove this theory. The parawing was supposed to be designed "so that even a nonpilot could fly." Meeker, on water skis, with the parawing overhead, was being dragged along behind a motorboat. The skis got caught underwater. Even though Meeker was a good water-skier, he was dragged, wet suit and all, underwater. He had trouble escaping from the heavy harness, and when he went under the cold waters of Lake Isabella,

his friends became frantic. Fortunately he survived this risky test (Leroy Riggs, personal communication, 23 August 1995).

15. Interview with Vice Adm. William Moran, 18 July 1988.

16. Wilcox, "Reminiscences," part 2, 14. Jagiello noted that the quotation might be accurate if it said McLean solved "most of the problems" (Howard Wilcox, comments on manuscript, 1995).

17. Telephone interview with Lee Jagiello, 22 July 1992; Otto J. Scott, *The Creative Ordeal: The Story of Raytheon* (New York: Atheneum, 1974).

18. See Glatt, *Demise of the Ballistics Division.*

19. L. A. "Pat" Hyland, *Call Me Pat* (Virginia Beach: Donning, 1993), 232; interview with Simon Ramo, 28 July 1991.

20. Amlie's recollection of the salaries for GS-12s in the Sidewinder period was $7,400 a year, and about $9,000 for GS-13s. The salaries for technical personnel in comparable skill grades at places like Hughes appeared to be about one and a half, or perhaps twice, these figures (interview with Thomas Amlie, 23 September 1990). Robert Blaise also recalled taking a 50 percent cut in "take-home pay" to join NOTS in 1951 from a position at General Motors.

21. Theodore F. Gautschi, *An Investigation of Research and Development Projects,* NOTS AdPub 112 (China Lake: U.S. Naval Ordnance Test Station, August 1962), 111.

22. Memo from Experimental Officer to Technical Director of NOTS, NP45-15, 8 October 1951, China Lake collection.

23. Gordon Thomas and Max Morgan-Witts, *Ruin from the Air: The Atomic Mission to Hiroshima* (London: Sphere, 1977), 371, 373. Interestingly, Capt. Frederic Ashworth, U.S. Navy, who later commanded China Lake, played a similar role on the Nagasaki mission; Parsons and Ashworth were the only two navy officers at Los Alamos. See Richard Rhodes, *The Making of the Atomic Bomb* (New York: Simon and Schuster, 1986), 739–40.

24. In Robert Blaise's papers is a letter from Lt. T. J. Christman, dated 17 August 1952, complimenting Blaise on his test plans: "I should like to compliment you on your thoroughness in writing up your test plans. When I read your plans, I am able to formulate immediately on what basis you are proceeding. . . . In a program in which many phases are difficult to tie down, it is a pleasure to be able to read your plans and know immediately in which direction you are going."

[288]

25. Interview with Rear Adm. Thomas J. Christman, 25 March 1993, 25, 9.

26. Thomas S. Amlie, letter to the author, 29 April 1988. The reference to "the only successful guided missile laboratory" was an exaggeration, of course; other projects, such as Sparrow III, were contenders. Sidewinder, however, was destined to be the star, so in this sense Amlie was correct.

27. Interviews with Douglas Wilcox, 28 February 1987, and Lee Jagiello, 18 July 1988.

28. Interview with Douglas Wilcox, 28 February 1987.

29. Interviews with Lee Jagiello and Vice Adm. William B. Moran, 18 July 1988; interview with Jacob Rabinow, 1 March 1988. Charles F. Woll of Philco also spoke long and enthusiastically to the chief of the Bureau of Ordnance, M. F. Schoeffel, about how easy Sidewinder would be to produce, "as easily as the simpler television sets" (Chief of the Bureau of Ordnance, letter to Commander, U.S. Naval Test Station, 25 September 1952). [289]

30. William B. McLean, "The Art of Simple and Reliable Design," spring 1963, publication source unknown, from *The Collected Speeches of William B. McLean,* NAWC Retrievable Manuscript RM-24 (China Lake: Naval Air Warfare Center Weapons Division, September 1993); emphasis added. Telephone interview with Emory Ellis, 14 December 1995. Frank Knemeyer noted in comments on the draft that an engineer had to go to the Gulf of Tonkin to uncover a key problem in the ELINT (electronic intelligence) operation.

31. Wilcox, "Reminiscences," part 4, 5; Frederick Ordway and R. C. Wakeford, eds., *International Missile and Spacecraft Guide* (New York: McGraw-Hill, 1960), 33.

32. Chief of the Bureau of Ordnance (M. F. Schoeffel), letter to Commander, U.S. Naval Test Station, Serial no. 78-1 (126) 25 September 1952, regarding Sidewinder.

Chapter 8. The Painted Bird

1. Interview with William Woodworth, 21 August 1992.

2. Henry Swift, letter to the author, 15 June 1992.

3. Some interesting insights on testing are contained in an article by Glenn Bugos, "Manufacturing Certainty: Testing and Program Management for the F-4 Phantom II," *Social Studies of Science* 23, no. 2 (May 1993): 265–300.

4. Letter to Newton Ward from "Van" (assumed to be Lt. Cdr. R. R. Vancil at the conclusion of his two-year tour as chief project pilot), Armitage Field, 30 August 1954, Newton Ward Papers.

5. Interview with Robert Blaise, 25 April 1991; interview with Frank Cartwright, 27 August 1991.

6. These remarks by an unnamed manager are mentioned in the interview with William McLean by Albert Christman, China Lake Oral History S-88, 16 November 1973, 12.

7. Interview with Robert Blaise, 15 April 1989 and 7 July 1991.

8. Intelligent redirection of funds, commonly known as "bootlegging," has been associated with the early stages of many critical weapons systems.

See Chalmers Sherwin, "Multiple Weapon System Advances through Multiple Innovations," in *Climate for Creativity,* ed. Calvin Taylor (New York: Pergamon, 1972), 143. See also Richard A. Muller, "Innovation and Scientific Funding," *Science* 209 (22 August 1980): 880–83, a rare admission by a bootlegger of how he used redirected funds to win a prestigious scientific award. Professor Muller later told me that after the article was published even his regular sources of funds dried up.

9. Capt. Walter M. Schirra Jr., *Schirra's Space* (Boston: Quinlan Press, 1988), 32–33.

10. Interview with Capt. Walter Schirra, 11 June 1990.

11. Interview with Capt. Thomas P. Rogers, 6 January 1995.

12. Interview with Robert Blaise, 25 March 1989, 6.

13. Robert Blaise, comments on draft, 25 April 1991.

14. Interview with Cdr. Glenn Tierney, 14 August, 1992, 54. Capt. Walter Schirra has questioned the authenticity of this story, saying the tone was already being used during the original tests before Tierney arrived.

15. In the mid-1990s VX-5 and VX-4 at Point Mugu merged to become VX-9.

16. "GMU-61 Has Vital Role in Sidewinder Program," *Rocketeer,* 22 March 1957, 4.

17. Interview with Cdr. Glenn Tierney, 14 August 1992, 82.

18. Interview with Cdr. Glenn Tierney, 14 August 1992, and later corrections.

19. Interview with Cdr. Glenn Tierney, 14 August 1992, 68.

20. Cdr. Glenn Tierney, letter to the author, 25 September 1994. Lt. Carl Quitmeyer also got assigned to China Lake through such interaction with Blackie Weinel (Carl Quitmeyer, letter to the author, 1 March 1996).

21. Cdr. Glenn Tierney, letter to the author, 25 September 1994.

22. Interview with Capt. Walter Schirra, 11 June 1990. This incident was seen differently by David Simmons and Tom Wong, who remembered the missile going straight up (David Simmons, interview, 22 February 1995).

23. In a memo written to Howard Wilcox, 8 July 1955, Tierney reports the two shots with dummy (no guidance) missiles after the fact and suggests that the success of the trial argues that more supersonic shots should be taken.

24. Robert B. Blaise, comments on manuscript.

25. Interview with Cdr. Glenn Tierney, 3 February 1995.

26. Robert B. Blaise, letter to the author, 1 April 1989; telephone interview, 25 March 1989; and comments on draft, 20 April 1991.

27. Telephone interviews with Robert Blaise, 25 March 1991, 30 May 1991, and 29 May 1991.

28. Telephone interview with Cdr. Glenn Tierney, 29 January 1993.

Chapter 9. Crunch Time

1. Interview with Edward Everett Benton Jr., 27 August 1991.
2. Interview with Leonard T. Jagiello by Leroy Doig III, China Lake Oral History S-168, 15 January 1988, 7. There is also evidence for the use of NOL White Oak wind tunnel on Sidewinder tests in the Rugg Notes.
3. Robert Blaise, letter to the author, 1 April 1989. He notes the contributions of Gene Curry and Jim Campbell to this effort.
4. Telephone interview with Warren Legler, 5 February 1989.
5. Robert Blaise, comments on draft, 25 April 1991.
6. Roderick McClung, Naval Weapons Center Oral History interview with David Rugg, S-198, circa 1970.
7. Interview with Roderick McClung by Elizabeth Babcock, 1 October 1991.
8. Interview with David Simmons, 20 February 1995.
9. Interview with Roderick McClung and David Simmons by David Rugg, China Lake Oral History Interview, S-198, circa 1970.
10. Interview with Roderick McClung, 26 August 1989. In the original Rugg interview, McClung mentioned half an hour. He later decided that half an hour was a bit of exaggeration and that two hours was probably more accurate as to developing time.
11. Frank Cartwright, in Sidewinder Oral History, 28 August 1991, 9.
12. Interview with Edwin G. Swann by David Rugg, about 1970.
13. Interview with Roderick McClung, 22 February 1995.
14. Interviews with Frederick Davis, 8 January 1988 and 28 July 1991.
15. Ralph Sawyer, letter to L. T. E. Thompson, 22 December 1952, L. T. E. Thompson Papers, China Lake.
16. Interview with Adm. Thomas Moorer, former head of the Joint Chiefs of Staff, 23 July 1991. Moorer was the experimental officer for NOTS 1951–53. Charles Townes, letter to the author, 3 January 1992. Professor Townes is well known for his invention of the maser.
17. The fear that weapon systems would be "taken" by a laboratory with more legitimate cognizance was well founded. Transferring a system from one lab to another raised the possibility that "not invented here" considerations would lead to half-hearted or malicious treatment, leading to the system's effective demise. This did in fact happen with Thomas G. Lang's SWATH ship, developed at one laboratory and moved to another (Thomas G. Lang, personal communication). Although the thought is continuous here, the remarks come from separate interviews with Robert Blaise on 25 March 1989 and 17 May 1990. Telephone interview with Frederick A. Darwin, 16 April 1990.
18. Interview with Robert Blaise, 1 May 1989.

[291]

19. Telephone interview with Peter Drucker, 15 July 1991; Howard Wilcox, "Research Problems Associated with the Development of Sidewinder," paper presented at a symposium sponsored by the Office of Naval Research as part of its Decennial Year, 19 and 20 March 1957. Other team members, such as Robert Blaise, considered the opposition threatening.

20. NOTS Technical Memorandum 1653, June 1953; interview with Rear Adm. Thomas J. Christman, 25 March 1993, 41.

21. Interview with Fred Davis, 8 January 1988.

22. This explanation was provided in a letter from Thomas S. Amlie to the author, 9 May 1989.

23. Walter Schirra remembers having fired this shot, but his flight log does not support his memory, and David Rugg's interviews with Nancy Carter and Lawrence Nichols seem to establish that it was indeed Albert Yesensky who fired the first "successful" shot. This was also the impression of Howard Wilcox, who remembers commiserating with Schirra about being unable to take the successful shot when so many others he had fired were unsuccessful. Fred Davis remembered the drone's tail being clipped (interview, 28 July 1991). But according to David Rugg's notes from NOTS TPR-84, the miss distance was eleven feet.

24. Interview with Frank Cartwright, 27 August 1991.

25. Rear Adm. W. S. Parsons, letter to Drs. F. Brown and W. B. McLean, 16 September 1953.

26. L. A. "Pat" Hyland, *Call Me Pat* (Virginia Beach: Donning, 1993), 338. McMath may have been part of the Research Development Board committee on guided missiles at this time.

27. Interview with Rear Adm. Thomas J. Christman, 25 March 1993, 14. The team checked with Lee Jagiello before removing the rollerons. Jagiello told them that aerodynamic roll damping would be sufficient to prevent roll at low altitudes (comments on manuscript by Lee Jagiello, December 1995).

28. Several people remember the 9 January shot as being the one just described. However, Rugg's notes cite technical progress reports no longer available to the effect that the shoot-down was first and the nonfatal damage was a later shot.

29. Warren Legler, letter to Howard Wilcox, 9 December 1988, 3.

30. Interview with Lawrence Nichols by David Rugg, about 1970.

31. Witnesses to this telegram include Howard Wilcox, Robert Blaise, and Capt. Walter Schirra.

32. Interview with Rear Adm. Thomas J. Christman, 25 March, 1993. Lt. Christman's diplomatic skills again proved of value to the project. No one had bothered to arrange dinner with McMath while he was visiting.

Lieutenant Christman immediately did so himself. Interview with Frederick Davis, 28 July 1991.

Chapter 10. To the Fleet

1. Jacob Rabinow, "The Individual in Government Research and Innovation," in *Innovation and U.S. Research: Problems and Recommendations*, ed. W. Norris Smith and Charles F. Larson (Washington, D.C.: American Chemical Society, 1980), 162. The other candidate he had in mind may have been Ted Toporeck, head of the Test Department. Shortly after McLean's promotion, Toporeck left for the Ramo-Wooldridge Corporation (interview with Robert B. Blaise, 5 September 1990, and entries in NOTS Code Directory for 1954 and 1955). [293]
2. Thomas S. Amlie remembers this specification as having been written by Charles P. Smith and himself (Thomas S. Amlie, letter to the author, 11 September 1995).
3. Wilcox, "Reminiscences," part 2, 18.
4. Thomas S. Amlie, letter to the author, 29 April 1988.
5. Interview with Howard Wilcox, 8 May 1988.
6. Thomas S. Amlie, letter to the author, 29 April 1988.
7. Interviews with Capt. Thomas S. Rogers, 6 January 1995 and 8 July 1990. Rogers fired an additional 25 shots when he was with the fleet, making a career total of 117. Cdr. Glenn Tierney, in a letter to the author (4 August 1995), thinks this may be a world record. Carl Quitmeyer, letter to the author, 22 June 1990.
8. Wilcox, "Reminiscences," part 2, 20.
9. Interview with Cdr. Glenn Tierney, 14 August 1992, 13.
10. Howard Wilcox, comments on draft.
11. Interview with Edward E. Benton by Ron Westrum, Elizabeth Babcock, and Mark Pahuta, Naval Weapons Center Oral History, S-193, 26 August 1991, 6.
12. Interview with Howard Wilcox, 1 March 1987.
13. Burton Klein, letter to Eric Gustafson, Harvard University, 2 March 1962.
14. Interview with Jacob Rabinow, 1 March 1988.
15. Wilcox, "Reminiscences," part 3, 7.
16. David Rugg note cards, in possession of the author.
17. Information on the BuOrd evaluation was given to me by Robert Blaise in a letter dated 30 May 1991. This account reflected Blaise's reconstruction of the test program in consultation with Rear Adm. Kenneth Wallace.
18. David Rugg notes. A special manual was written for this evaluation: Lt. Rufo Robinson, "Preliminary Pilot's Handbook on Sidewinder XAAM-N-seven for Pilots of VX-3," China Lake, California, April 1955. This was

not the first contact between VX-3 and Sidewinder. Earlier, in May 1955, a "limited non-firing evaluation" by VX-3 had taken place to determine how Sidewinder might change fleet tactics.

19. Interview with Edward E. Benton Jr., 26 August 1991.

20. Thomas S. Amlie, comments on manuscript; interview with Robert B. Blaise, 11 July 1991.

21. Telephone interview with Robert B. Blaise, 8 June 1991.

22. David Rugg note cards (for 1956), in possession of the author.

23. Interview with Peter Nicol, 8 January 1988.

24. Interview with Peter and Polly Nicol by Elizabeth Babcock, William R. Blanc, Leroy Doig III, and Mark Pahuta, NAWCWPNDIV, interview number S-225, 17 March 1993. According to LaV McLean, Nicol was a perfectionist (interview with LaVerne McLean, 21 August 1995).

25. Sparrow I had been sent to the fleet before it was ready, leading to very unhappy experiences by naval aviators and ships' crews alike (interview with Max White, 17 June 1995).

26. Interview with Peter and Polly Nicol, 17 March 1993, S-225.

27. Interview with Peter Nicol, 8 January 1988.

28. Interview with Peter and Polly Nicol, 17 March 1993, 48.

29. Nicol's effectiveness was so widely recognized that he was asked for by name in official requests. One message describing a forthcoming Fleet Missile Operational Readiness Test included the line, "The services of your Mr. Peter Nicol are particularly desired." He received a Sustained Superior Performance Award from the navy for work done from 1958 to 1963.

30. Interview with Frank Wentink by David Rugg, about 1970. Making training films is part of what John Law has called "heterogeneous engineering," developing the human as well as the material components of a technology (Law, "Technology and Heterogeneous Engineering: The Case of the Portuguese Expansion," in *The Social Construction of Technological Systems,* ed. W. Bijker, T. Hughes, and T. Pinch [Cambridge, Mass.: MIT Press, 1987], 113).

31. Sidewinder designations are confusing. In general, I use current designations for the missiles, using the "AIM-9" prefix. AIM stands for "air intercept missile." Robert L. Lawson, ed., *The History of U.S. Naval Air Power* (New York: Military Press, 1987), 133, 148.

32. In June 1958 Lt. Carl Quitmeyer, just graduated from UCLA, was ordered to join the staff of Admiral Pride, commander Naval Air Forces, Pacific Fleet. Lt. Quitmeyer's staff role was to write the Sidewinder and Sparrow III introduction, training, firing, and deployment guidelines for the Pacific area squadrons. This billet reflected Quitmeyer's GMU-61 activities in testing the missile (see chap. 12; Carl Quitmeyer, letter to the author, 1 March 1996).

33. Frank Wentink interview by David Rugg, about 1970. The "foolproof" impression did not survive Sidewinder's combat experience.
34. Memorandum, "A Proposed Citation for Dr. William B. McLean," 13 October 1956 from F. S. Withington, Chief of the Bureau of Ordnance, in LaV McLean Papers.
35. U.S. Navy chart, "Cost Reduction Due to Value Engineering: Competition and Quantities for Sidewinder 1 and 1A Guidance and Control Contracts," about 1965.
36. Naval Weapons Center Annual Performance Report for December 1982.
37. The test pilot for the Sparrow III, Lt. Clyde "Andy" Andrews, played a major role in convincing McDonnell to design the plane around the Sparrow III (interview with Max White, 17 June 1995).

Chapter 11. Selling the Air Force

1. Unless otherwise indicated, all information in this chapter relating to Colonel Scheller's activities comes from a telephone interview with him on 27 March 1990.
2. Interview with Vice Adm. William Moran and Lee Jagiello, 18 July 1988.
3. Cdr. Glenn Tierney, letter to the author, 30 April 1994; interview with Cdr. Wade Cone, 19 July 1988.
4. Interview with Robert Blaise, Lt. Carl Quitmeyer, and Capt. Thomas S. Rogers, 21 July 1991.
5. Wilcox, "Reminiscences," part 3, 8–9.
6. Memorandum from G. G. Neice, Aide, to Capt. P. D. Stroop, Commanding Officer of NOTS, 11 August 1953.
7. Wilcox, "Reminiscences," part 3, 9. The report-cooking scenario Gardner proposed mirrors his methods of operation in developing the ICBM, where he stacked more than one committee; see Simon Ramo, *The Business of Science: Winning and Losing in the High-Tech Age* (New York: Hill and Wang, 1988), 109. See also Donald MacKenzie, *Inventing Accuracy: A Historical Sociology of Nuclear Missile Guidance* (Cambridge, Mass.: MIT Press, 1990), 110.
8. Interview with Thomas S. Amlie, 6 September 1994. Elizabeth Babcock has questioned this story, saying that cloud cover at China Lake is slight.
9. Interview with Lt. Col. Thomas McElmurry, 4 January 1994, and Lt. Col. John Prodan, 13 January 1994.
10. Interview with Cdr. Glenn Tierney, 14 August 1992, 43. This account is contested by Lt. Col. John Prodan, at that time Falcon project officer (interview, 13 January 1994).
11. Interview with Cdr. Glenn Tierney, 14 August, 1992.
12. Charles F. Smith, revisions to interview of 27 March 1987.

13. Sources disagree on whether the "tailpipe shot" took place first or second; Benton, Blaise, Smith, and Wilcox (who was not present) put the tailpipe shot first. Tierney, Benton, and Blaise were there but disagree on details.

14. Blaise and many others on the Sidewinder project depended heavily on informal connections to make things happen. John Kelly in BuOrd was particularly helpful in coming up with "small pots of money" for things that needed doing but were too urgent to go through channels (interview with Robert B. Blaise, 10 July 1991).

15. There may have been other bets placed. Capt. Thomas S. Rogers remembers hearing that the air force had lost bets worth eight hundred dollars on the Holloman tests (interview, 8 July 1990).

16. Interview with Cdr. Glenn Tierney, 14 August 1992, 38.

17. Wilcox suspects that these new tests were designed to provide an impossible task for Sidewinder. His hypothesis is plausible but hard to check.

    The following account is based on the memories of Frank Cartwright, who was present for the tests, and Robert Blaise, who debriefed him shortly thereafter.

18. Cdr. Glenn Tierney, letter to the author, 25 September 1994.

19. Glenn Tierney had actually fired a Sidewinder in a supersonic dive earlier, "just to see what would happen." Interview with Cdr. Glenn Tierney, 14 August 1992.

20. Interview with Col. Donald Scheller, 2 March 1990. Scheller remembers this as being a multistanza song to the tune of "Davy Crockett" for the party that night, a farewell to the base's commander, Capt. Davy Young. However, Young left in June and the test was in September, so the song may have been another.

21. Interview with Frank Cartwright, 28 August 1991.

22. Lee Jagiello interview with Leroy Doig III, 15 January 1988, 10.

23. Interview with Thomas S. Amlie, 6 September 1994.

24. Commander, U.S. Naval Ordnance Test Station, letter to Chief, Bureau of Ordnance, Serial No. 112, 19 December 1956.

25. Warren Legler, letter to Howard Wilcox, 9 December 1988.

26. Travel clearance papers, Warren Legler.

Chapter 12. High-Altitude Testing

1. Telephone interview with Capt. Thomas S. Rogers, 8 July 1992.

2. Unless otherwise indicated, all information in this chapter relating to Glenn Tierney's activities comes from an interview with him on 14 August 1992.

3. L. A. "Pat" Hyland, *Call Me Pat* (Virginia Beach: Donning, 1993), 271.

4. Hyland, *Call Me Pat,* 272. James Gilkerson, Hughes Public Relations head at the missile operations in Tucson, was unable to find any trace of this event. However, a friend of Robert Sizemore told him a story quite similar to this one, so I am disposed to accept Hyland's account (interview with Robert R. Sizemore, 22 February 1994).
5. Interview with Robert Blaise, 25 April 1991. Howard Wilcox thinks the project at this point was much more secure than Blaise remembers (comments on draft).
6. This ability to move rapidly ("can do!") would also be shown when the time came to get Sidewinders quickly to the Chinese Nationalist forces during the Quemoy-Matsu crisis. See chapter 16.
7. Flight Log of Glenn Tierney, entries for 20–25 May 1956. The exact sequence here has been constructed from the memories of Tierney and Robert Blaise. Tierney's flight log shows his first flight in the F-100C, tail number 794, on 27 April 1956. The crushed Cannon plug problem is as recounted by Robert B. Blaise, 25 March 1989, 6.
8. The effective firing envelope typically consists of a minimum and maximum range, and a maximum angle off the target, which dictates how sharp a turn the missile will have to make to intercept its target. For the Sidewinder, it might be useful to think of it as a cone with the point at the target's tail.
9. Telephone interview with Robert R. Sizemore, 22 February 1994.
10. Interview with Cdr. Glenn Tierney, 14 August 1992, 40–41.
11. Robert Sizemore, in a telephone conversation, 21 July 1996, mentioned that it was his impression that up to that time no infrared Falcon had yet hit a Matador after some thirty tries.
12. Ray Wagner, *Modern American Combat Planes* (Garden City, N.Y.: Doubleday, 1982), 466. Actually the Falcon was designed for the F-106, but the F-106 (the "1954" interceptor) was nowhere near on schedule, so the F-102 was put in its place as an interim solution.
13. Interviews with Cdr. Glenn Tierney, 14 August 1992, and Robert B. Blaise, 25 April 1991.
14. Robert B. Blaise comments on draft, 25 April 1991.
15. Chuck Yeager and Leo Janos, *Yeager: An Autobiography* (New York: Bantam), 317.
16. Cdr. Glenn Tierney, letter to the author, 30 April 1994.
17. Interview with Cdr. Glenn Tierney, 14 August 1992, 20.
18. Interview with Cdr. Glenn Tierney, 14 August 1992.
19. Carl Quitmeyer, letter to the author, 1 March, 1996.
20. Robert Blaise, letter to the author, 1 April 1989, and article by Betty Campbell, "Submarine Deck Hangar Comes to Desert: Will Serve as Pyrotechnics Chamber," *Rocketeer,* 14 February 1958, 4.

21. Cdr. Glenn Tierney, letter to the author, 25 September 1994. According to Capt. Tom Rogers, GMU-61 wanted one of the navy's high-flying F-8 Crusaders as early as possible. It proved "a giant bureaucratic hassle" to obtain one, however, so tests with the air force's F-104 preceded those with the navy plane (interview with Capt. Thomas Rogers, 8 July 1990).

22. Interview with Col. Donald Scheller, 18 April 1990.

23. Interviews with Robert Blaise, 15 April 1989, and Thomas McElmurry, 10 January 1994; Lee Jagiello, comments on draft, December 1995.

24. Bert Kinzey, *F-104 Starfighter* (Blue Ridge Summit, Pa.: Tab Books, 1991), 17; telephone interview with Lt. Col. Thomas McElmurry, 9 January 1994. Operational use of the F-104 cost the Luftwaffe at least 160 crashes by 1974. This caused a national scandal in Germany. See Bill Gunston, *Early Supersonic Fighters of the West* (New York: Scribner's, 1976), 210.

25. Interview with Vice Adm. William Moran and Lee Jagiello, 18 July 1988, Moran speaking.

26. Interview with Col. Donald Scheller and Lawrence Nichols, 2 August 1992; telephone interview with Robert R. Sizemore, 22 February 1994.

27. Commander, NOTS, letter to Chief, Bureau of Ordnance, Serial 00112, 19 December 1956; interview with Lee Jagiello by Leroy Doig III, 15 January 1988, China Lake Oral History S-168, 10.

28. The incident related here is based on the memory of Cdr. Glenn Tierney, interview 14 August 1992, with his revisions.

29. U.S. Navy Bureau of Weapons, "Significant Accomplishment with the Advanced Sidewinder 1C (IR)," *Special Report for the White House,* 9 July 1963.

30. Glenn Tierney, orders from ComNOTS to proceed to Point Mugu for guided missile demonstration, U.S. Naval Ordnance Test Station, Serial T-114-56, 6 January 1956.

31. Research Board notes from 12 January 1956 show "Capt. Ashworth reported that Station officials had travelled to Point Mugu the previous day to be present at briefings for Adm. Burke. . . . Three Sidewinder missiles were fired at F6F drones, all of the missiles homed and the warheads detonated, but the third round damaged the drone target. The shots were considered 'successful, but not spectacular.' Six Sparrow missiles were fired, one of which destroyed the drone with warhead detonation" (Elizabeth Babcock, letter).

32. Robert Blaise, "The Theory and Practice of Creative Environments," typescript, 25 August 1991, 12.

33. Edwin Swann remembers Hugh D. Woodier and Paul Cordle as being key members of the team that developed the expanding rod warhead (Edwin G. Swann Jr., China Lake Oral History Interview with David Rugg, S-198, circa 1970, B-25).

34. Thomas S. Amlie argues that the expanding rod warhead was not that much of an improvement and notes that the AIM-9L has a far superior blast fragmentation warhead (letter to the author, 11 September 1994).
35. Historical Branch, Air Force Missile Development Center, U.S. Air Force, *Contributions of Balloon Operations to Research and Development at the Air Force Missile Development Center, 1947–1958,* Holloman, N. Mexico, n.d. (but probably early 1959). A copy of this document was obtained from the Los Angeles Public Library.
36. Telephone interview with Lt. Col. Thomas McElmurry, 4 December 1994.
37. Historical Branch, *Balloon Operations,* 50, 51.
38. Richard Vogt had been chief engineer for the Blohm and Voss aircraft company in Germany and was famous for his successful design of an aircraft with asymmetrical wings.
39. Interview with Capt. Thomas S. Rogers, 23 July 1992.

Chapter 13. The Creative Dialogue

1. Interview with Jacob Rabinow, 5 July 1991. Rabinow noted that Seymour Cray, the computer designer, often stated that any computer design team greater than thirty-five people was too large. See also R. B. Kershner, "The Size of Research and Engineering Teams," *I.E.E.E. Transactions on Engineering Management* (June 1958): 35–38. McLean's speeches are available from the Technical Information Department at China Lake as *Collected Speeches of William B. McLean,* NAWC Retrievable Manuscript RM-24, September 1993.
2. McLean stated, in a letter of 8 June 1961, to Dr. Clifford C. Furnas, chancellor, the University of Buffalo, that he considered Charles J. Hitch and Roland N. McKean's chapter 13 "Military Research and Development" in their book *The Economics of Defense in the Nuclear Age,* 2d ed. (New York: Atheneum, 1967) as an excellent statement of the major problems: "This chapter very clearly expresses the things which I feel I have learned and experienced during 20 years of work on the development of new military equipment." The book was originally published in 1960 by the RAND Corporation.
3. William B. McLean, "Operation of Navy Laboratories in a Society Dominated by Technological Progress," speech delivered in March 1960 for the Tenth Annual Meeting of Senior Scientists' Council of Navy Laboratories, NOL Corona, California.
4. William B. McLean, letter to Dr. Clifford C. Furnas, chancellor, the University of Buffalo, 8 June 1961.
5. William B. McLean, "The Sidewinder Missile Program," in *Science, Technology, and Management,* ed. Fremont Kast and James Rosenzweig (New York: McGraw-Hill, 1963). The paper originally was delivered at

the National Advanced Management Conference in Seattle, Washington, 3–7 September 1962.

6. William B. McLean, letter to Vice Adm. John T. Hayward, 15 June 1959. No doubt Vice Admiral Hayward, a highly respected former experimental officer at China Lake, was sympathetic. Actually, in the early sixties, China Lake did persuade the navy to make special arrangements to avoid this fragmentation of responsibility (interview with Franklin Knemeyer, 23 August 1995).

7. William B. McLean, "Management and the Creative Scientist" in *California Management Review* 3, no. 1 (fall 1960): 9–11. This was delivered originally to a conference entitled "It Depends on Where You Sit." Henry Swift indicated in a letter to the author (6 May 1991) that this article was written and published as originally dictated by McLean.

8. On the AR-15, see Thomas McNaugher, *The M16 Controversies: Military Organization and Weapons Acquisition* (New York: Praeger, 1984); see also Gregg Easterbrook on the F-20 Tigershark fighter's rejection by the air force, "The Airplane That Doesn't Cost Enough," *Atlantic Monthly*, August 1984, 46–56.

9. Interview with Vice Adm. William J. Moran by Leroy Doig III, 5 December 1990, China Lake Oral History S-187, 53; emphasis added. Likewise, at the Lockheed Skunk Works, Kelly Johnson had an air force project office on site so there would be no delay and no breakdown in communication (Clarence L. "Kelly" Johnson with Maggie Smith, *Kelly: More Than My Share of It All* [Washington, D.C.: Smithsonian Institution Press, 1985]).

10. The classic example of this problem is the F-111B fighter-bomber, whose early, "concurrent" production strategy caused the wasteful production of some four hundred planes that were nonfunctional. It probably cost $2 billion to produce these planes, essentially scrap metal from the day they were built. See Robert F. Coulam, *Illusions of Choice: The F-111 and the Problem of Weapons Acquisition Reform* (Princeton, N.J.: Princeton University Press, 1977).

11. See, for example, Chalmers Sherwin and R. S. Isenson, "Project Hindsight," *Science* (June 1967).

12. Amrom Katz, personal communication. See also Burton Klein "The Decision-Making Problem in Development," in *The Rate and Direction of Inventive Activity: Economic and Social Factors, National Bureau of Economic Research* (Princeton, N.J.: Princeton University Press, 1962), 477–97.

[300]

Chapter 14. Early Generations

1. NOTS Memo 352/WBL:la, /a1-1(567), Serial 01786, 20 November, from ComNOTSs to Chief BuOrd, NOTS Sidewinder Speedletter no. 12, 1–2.

2. NOTS Memo 352/WBL:la, /a1-1(567), Serial 01786, 20 November, from ComNOTS to Chief BuOrd, Speedletter no. 12. See also Naval Weapons Center Technical Paper 6413, part 1, 32.
3. Elizabeth Babcock, "History of Guided Missiles," China Lake working document, undated; Ray Wagner, *American Combat Planes,* 3d ed. (Garden City, N.Y.: Doubleday, 1982), 521.
4. NOTS Speedletter no. 12, 4.
5. Bill Gunston, *The Illustrated Encyclopedia of the World's Rockets and Missiles* (London: Salamander Books, 1979), 226.
6. Telephone interview with John Joyner, 10 October 1995.
7. Interview with Ken Powers by Elizabeth Babcock, China Lake Oral history, 18 May 1981, 7.
8. Walter J. Boyne, "New Mexico Nightmare," *RAF Flying Review* 17, no. 7 (June 1962): 15–17, quotation on 17. Most of this section is based on Boyne's article.
9. Telephone interview with Robert B. Blaise, 25 April 1991.
10. Interview with Cdr. Glenn Tierney, 14 August 1992.
11. Robert Blaise, comments on draft, 25 April 1991.
12. The missile could pull 20 Gs, far beyond any human pilot's ability. Over North Vietnam, however, North Vietnamese pilots did sometimes escape by going into afterburner and turning hard into the missile. They did not out-turn the missile, but they changed the geometry of the encounter enough to defeat it (based on comments by Richard Schmitt on draft).
13. Frank Knemeyer, comments on draft. Development of the cooler cost $400,000.
14. Thomas S. Amlie, letter to the author, 5 October 1994. The radar homing AIM-9C worked equally well and allowed head-on shots (which the AIM-9D's infrared system didn't). But when its mated aircraft—the F-8 Crusader—left the fleet, it did too, and thus eventually left the navy's inventory (Bill Gunston, *Modern Airborne Missiles* [New York: Prentice-Hall Press, 1986], 54). Tom Gervasi, *Arsenal of Democracy: American Weapons Available for Export* (New York: Grove Press, 1977).
15. Charles P. Smith, in a telephone interview, 19 May 1996, indicated that his own view of Philco's work was more positive, that they experienced problems but nothing beyond the expected in such innovative efforts.
16. Oral history interview with Ken Powers by Elizabeth Babcock, 18 May 1981, 3.
17. Thomas S. Amlie, letter to the author, 19 October 1989.
18. The materials in this section are derived largely from two interviews with Burrell Hays, 8 January 1988 and 20 March 1994.
19. Interview with Joseph D. Pasquale, 23 February 1995.

[301]

20. Interview with Burrell Hays, 20 February 1995.
21. Philip J. Klass, "Competition Slashes Sidewinder 1A Price," *Aviation Week and Space Technology,* 9 April 1962, 91–95; interview with Capt. Thomas Rogers, 23 July 1992. Second sourcing works because the "learning curve" slopes more steeply when firms are trying to cut costs, so the price per unit drops rapidly. Cost is lowered by more efficient production and by substituting less expensive materials and subassemblies. On the Sidewinder AIM-9B, for instance, the price in 1962 dropped to one-third of the 1956 price, thanks to competition (David Rugg notes). See also Frederic M. Scherer, *The Weapons Acquisition Process: Economic Incentives* (Boston: Harvard University Graduate School of Business Administration, 1964), 117–28. Second sourcing was used extensively in World War II.
22. One example of a quality control program: The China Lake solder specification became famous. Contractor personnel would come to the Naval Weapons Center by the hundreds, using the largest auditorium on the base, to be trained how to meet the specification. Burrell Hays later marveled that contractor personnel from CEOs on down came to China Lake to take these seminars. Some had to stay nearly two hours away, four to a room, in Palmdale and yet arrive at NOTS by 7 A.M. Whatever their feeling, they did it (interview with Burrell Hays, 20 February 1995).
23. Telephone interview with Charles P. Smith, 8 May 1996.
24. According to Frank Knemeyer and Eddie Allan, there was a pilot production line that produced 130 rounds used for operational evaluation.
25. Telephone interview with Richard Beckerleg, 17 October 1996. Edward E. Benton (telephone interview 19 October 1996) confirmed the difficulties with getting parts to fit.
26. Interview with Norman Woodall, 25 February 1996; interview with Burrell Hays, 8 January 1988. Richard Beckerleg and Edward Paul, both formerly with Raytheon, think the big problems were with the data package and not with Raytheon's manufacturing methods. Richard Beckerleg commented, "You can specify things on a drawing, but if you can't really do them, then things have to be changed. . . . There were things that China Lake didn't know would make a difference, but they did" (telephone interview with Richard Beckerleg, 17 October 1996).
27. Interview with Burrell Hays, 20 February 1995; telephone interview with Charles P. Smith, 14 March 1996.
28. The friendly coordination between China Lake and Raytheon on the design side was counterbalanced by a constant tug-of-war on the production end. Raytheon constantly challenged China Lake by its attempts to cut corners in the production of Sidewinder. When China Lake got

sloppy in its surveillance, Raytheon took advantage (interview with Lee Sutton, 5 November 1993).

29. Most of the information for this section comes from David Rugg's notes, written in the early 1970s from interviews. Due to absence of the documents on which these notes were based, few could be checked.

30. Thomas S. Amlie, letter to the author, 26 October 1994; telephone interview with John Boyle, 28 September 1994.

31. The supersonic Chance-Vought F8U-2N Crusader was a pure air-superiority fighter. See Barrett Tillman, *MiG Master: The Story of the F-8 Crusader*, 2d ed. (Annapolis, Md.: Naval Institute Press, 1990), and the even better Paul T. Gillchrist, *Crusader! Last of the Gunfighters* (Atglen, Pa.: Schiffer Publishing, 1995). See also George Hall, *Top Gun: The Navy's Fighter Weapons School* (Novato, Calif.: Presidio Press, 1987), 124.

32. Telephone interview with Earl Donaldson, 6 February 1994.

33. Thomas S. Amlie admitted in a letter to the author that the idea for the radar seeker itself had originally come from Hughes. It was one of the ideas that Hughes engineers had considered but passed over. Obviously "not invented here" did not concern Amlie (Thomas S. Amlie, 8 March 1996).

34. Thomas S. Amlie, Trip Report, 354/TSA:drq, Serial 26, 13 May 1960.

35. Thomas Amlie, letter to the author, 26 October 1994. Nonetheless, Raytheon would soon be manufacturing Sidewinder guidance and control sections for the AIM-9D and AIM-9G (Raytheon Sidewinder chronology). Charles P. Smith, letter to the author, 13 June 1996.

36. David Rugg interview with Frank Wentink, about 1970.

37. Thomas S. Amlie notes that Motorola, which manufactured the seeker, did an excellent job, and that after a few initial quality problems, "everything they sent me was Tiffany jewelry" (Amlie, letter to the author, 11 September 1994).

38. Thomas S. Amlie, letter to the author, 26 October 1994.

39. Chuck Yeager and Leo Janos, *Yeager: An Autobiography* (New York: Bantam, 1985), 313.

40. This information was generated both by airborne warning and command system (AWACS) aircraft and also by an ability to interrogate the North Vietnamese IFF, thus clearly identifying attackers as hostiles.

41. Telephone interview with Charles P. Smith, 14 March 1996.

42. Marshall Michel, *Clashes: Air Combat over North Vietnam, 1965–1972* (Annapolis, Md.: Naval Institute Press, 1998), 228.

43. Walter Freitag pointed out that whereas vacuum tubes require a high-voltage, low-current environment, solid-state circuits need a low-voltage, high-current environment (telephone interview with Walter Freitag, 19 June 1996). See also memo from Head of Electronic Circuit Design

[303]

Section to SW/1C Project Office, "Transistorization of the Sidewinder and Chaparral GCG's," Reg. 5522-911, 17 December 1965.

44. When Raytheon got the production contract, however, they used stiff boards instead of the flexible harnesses (interview with Walter Freitag, 19 June 1996).

45. Telephone interview with Norman Woodall, 25 February 1996. Israel got the earlier models for its Yom Kippur War. Some 220 AIM-9Ds and 480 AIM-9Gs were rushed to Israel during the crisis and supposedly got a kill ratio *per engagement* of 92 percent (Gervasi, *Arsenal of Democracy,* 180).

46. This section barely scratches the surface of Sidewinder variants. In addition to air-to-ground versions such as Focus, Sidewinder seekers were put on a variety of airframes, as indicated by names like Sparrowinder, Bullwinder, and so on. There were also subwinders, spacewinders, and so on. Sometimes it seemed that every conceivable combination of seeker, airframe, and launch regime was tried at least once.

47. Charles P. Smith, letter to the author, 13 June 1996.

48. Charles P. Smith, letter to the author, 13 June 1996.

49. Gunston, *Illustrated Encyclopedia,* 176.

Chapter 15. Later Generations

1. Telephone interview with Charles P. Smith, 20 December 1995.

2. Telephone interview with Ed Paul, 1 June 1997.

3. Actually, there was a significant change from the first to the second generation of Sidewinders. While the optical bandpass for the AIM-9B was 2 to 2.5 microns, AIM-9D with its cooled PbS detector was 3.2 to 5 microns. This allowed the AIM-9D to receive about 35 percent of the available IR signal, while AIM-9B got only 7 percent of the IR signal (Charles P. Smith, letter to the author, 13 June 1996).

4. Interview with Edward E. Benton by Ron Westrum and Elizabeth Babcock, NWC Oral History Interview S-193, 26 August 1991.

5. The "rate bias" system moved the missile's aim point to the farthest forward "hot spot" on the target. Rate bias had been developed by Hughes to solve the "friendly Falcon" problem that caused the missile to miss in Vietnam. Raytheon then picked this concept up and effectively applied it so that Sidewinders went after the jets and not the exhaust plumes (Richard Schmitt, comments on draft, 1997; telephone interview with Edward Paul, 1 June 1997).

6. Interview with Richard Schmitt, 24 August 1995.

7. Telephone interview with Louis Covert, 18 July 1996.

8. According to Walter Freitag, the prime designers were Louis Covert, Richard Schmitt, Vern Anderson, and Bert van den Berg (telephone interview with Walter Freitag, 19 June 1996).

9. Interview with Richard Schmitt, 24 August 1995. Schmitt pointed out that a decision about a single resistor might entail a thirty-thousand-dollar difference in production costs.

10. Telephone interviews with Ed Paul, 31 October 1996 and 1 June 1997. It was also known from Vietnam that the 9J was a very inferior missile to the navy's 9G (Marshall Michel, *Clashes: Air Combat over North Vietnam* [Annapolis, Md.: Naval Institute Press, 1997], 228).

11. Telephone interview with Edward Paul, 16 November 1996. [305]

12. Telephone interview with Ernest Cozzens, 10 May 1996.

13. Telephone interview with Lawrence Washam, 19 January 1997. Another test pilot, Lt. C. C."Andy" Andrews, had helped convince McDonald Douglas to accept the Sparrow III (interview with Max White).

14. Russell Warren Howe, *Weapons: The International Game of Arms, Money, and Diplomacy* (Garden City, N.Y.: Doubleday, 1980), 156. See also Bill Gunston, *The Illustrated Encyclopedia of the World's Rockets and Missiles* (London: Salamander Books, 1979), 213. Telephone interview with Norman Woodall, 25 February 1996.

15. Entry "Sidewinder (AIM-9L)" in *Jane's Weapons Systems 1978* (London: MacDonald and Janes, 1977). Norman Woodall credits Raytheon with solving most of the reliability problems with the missile (telephone interview, 25 February 1996).

16. Interview with Thomas S. Amlie, 6 September 1994; telephone interview with Russell Whynot, 15 May 1997.

17. Interview with Richard Schmitt, 24 August 1995; telephone interview with Capt. Willard Van Dyke, 20 January 1997.

18. Telephone interview with Norman Woodall, 25 February 1996; Capt. Willard Van Dyke, in his telephone interview 20 January 1997, agreed.

19. Radar missiles had far superior detection range to anything infrared could manage but did not provide very good resolution of the target. Infrared could provide better resolution.

20. Telephone interview with Edward Paul, 1 June 1997. The departure of Burrell Hays from the position of technical director also was a factor in allowing this situation to occur.

21. Apparently the navy contract allowed for several major hardware or software changes that could be requested for no additional funds. It is no surprise that Raytheon top management felt that signing on this contract would be financial suicide (interview with Edward Paul, 8 October 1996).

22. Telephone interview with Capt. Jess Stewart, 2 November 1996; telephone interview with Billy Boatright, 14 June 1996. The late Commander Boatright was formerly division director of Air-to-Air Missile Projects at China Lake.

23. Don Flamm, "Sidewinder Contract Zooms Along after 18 Years, $950 Million," *Orange County Businessweek,* 19 October 1987, 4, 5.

24. Telephone interview with Louis Covert, 18 June 1996.

25. Telephone interview with Billy Boatright, 14 June 1996.

26. Interview with Madelyn Fortune, 22 February 1995.

27. Telephone interview with Edward Paul, 16 November 1996; telephone interview with David Kurdeka, 10 June 1996.

28. Telephone interview with Michelle Bailey, 22 July 1997.

29. Telephone interview with Louis Covert, 18 June 1996.

30. "China Lake Readies Data for 1990's Sidewinder Upgrade," *Aviation Week and Space Technology,* 20 January 1986, 80–81. Capt. Jess Stewart, who was sympathetic to China Lake's work on the AIM-9R, was replaced just at this time with a successor whom he felt did not understand the value of what China Lake was doing (telephone interview with Captain Stewart, 2 November 1996).

31. Norman Woodall, letter to the author, 21 May 1996.

32. "Sidewinder Team Finds Right Recipe for Successful AIM-9M(R) Project," *Rocketeer,* 46, no. 13 (4 April 1991); also Norman Woodall, letter to the author, 21 May 1996.

33. Key contributors on this team included Bill Walters and Dan Crabtree (Richard Schmitt, comments on draft, 1997).

34. David Hughes, "Off-Boresight Missile Test Highlights U.S. Capability," *Aviation Week and Space Technology,* 6 June 1994, 66–69.

35. Apparently a defecting Russian pilot informed the Americans that the USSR assumed the United States was developing Agile covertly.

36. Michael Dornheim and David Hughes, "U.S. Intensifies Efforts to Meet Missile Threat," *Aviation Week and Space Technology,* 16 October 1995, 36–39.

37. This was still obvious in 1996, according to information put on the Internet by the Joint Navy and Air Force Sidewinder Team: "Air Intercept Missile-9X (AIM-9X) Sidewinder: FY96 Activity." According to an article in *Interavia,* ASRAAM is expected to be in production in 1998; see Nick Cook, "ASRAAM Leads UK Missile Revolution," *Interavia,* August–September 1996, 39.

38. Bill Gunston, *Modern Airborne Missiles* (New York: Prentice Hall, 1986), 36, 50.

39. Art Hanley, "China Lake," *Aerospace America,* January 1991, 8–10.

40. Telephone interview with Charles P. Smith, 8 May 1996.

41. Unknown Soviet author, translated manuscript, 110; emphasis in original.

42. Interview with Fred Davis, 26 August 1992; Gunston, *Modern Airborne Missiles,* 12.

Chapter 16. In Combat

1. This book has benefited enormously from the excellent work by Marshall Michel, *Clashes: Air Combat over North Vietnam, 1965–1972* (Annapolis, Md.: Naval Institute Press, 1997). The Tierney quote is from Glenn Stackhouse, "Navy Unveils Deadly Air-to-Air Missile," *San Jose Evening News,* 4 March 1957, 12.

2. Robert W. Love Jr,. *History of the United States Navy,* vol. 2 (Harrisburg, Pa.: Stackpole, 1992), 428.

3. Interview with Robert Sizemore, 22 February 1994.

4. U.S. Office of Naval Intelligence, Information Reports, (Source: ALUSNA) Report 45-5-58, Sitrep no. 19, 26 September 1958.

5. Michel, *Clashes,* 287. The Sparrow figures reflect the missiles fired. Many more were never even used because they were defective.

6. Richard Schmitt, in comments on the draft in 1997, suggested that Falcon had done well in testing when used against small drones. When fired in combat against larger jets, Falcon went for the plume, thus giving it the "friendly Falcon" moniker. Hughes finally did fix the problem, but too late for the Vietnam War.

   F. S. Withington, "Detailed Specification for the Recommendation of Dr. William B. McLean for a Presidential Citation and Monetary Award," 13 October 1956, William B. McLean Papers, in the home of LaVerne McLean.

7. Interview with Burrell Hays, 20 March 1994; telephone interview with Thomas S. Amlie, 8 March 1996. To cope with such influence, the commanding officer of VX-4 had a yellow line painted on the floor of the missile shop and contractors were not allowed to cross it. When the AIM-9D and AIM-9C were undergoing operational evaluation, the same rules applied to the China Lake team, even though they were government employees (Thomas S. Amlie, personal communication, 10 June 1995).

8. Interview with Lt. Col. John Prodan, 13 January 1994.

9. John B. Nichols and Barrett Tillman, *On Yankee Station: The Naval Air War over Vietnam* (Annapolis, Md.: Naval Institute Press, 1987), 72–73.

10. Michel, *Clashes,* 59–60.

11. Robert K. Wilcox, *Scream of Eagles* (New York: John Wiley, 1990).

12. Walter J. Boyne, *Phantom in Combat* (Washington, D.C.: Smithsonian Institution Press, 1985), 46.

13. Wilcox, *Scream of Eagles,* 96–97.

14. Quoted in Paul T. Gillchrist, *Crusader! Last of the Gunfighters* (Atglen, Pa.: Schiffer Publications, 1995), 192.

15. Michel, *Clashes,* 278–79.

16. For a complete account, see Wilcox, *Scream of Eagles.*

17. Capt. Frank Ault, "Ault Report Revisited," *The Hook* (spring 1989): 36.

[307]

18. Naval Air Systems Command, *Report of the Air-to-Air Missile System Capability Review,* 1 January 1969.
19. Wilcox, *Scream of Eagles.*
20. See for instance, Randy Cunningham with Jeffrey Ethell, *Fox Two* (New York: Warner Books, 1989).
21. Michel, *Clashes,* 278, 288–91.
22. Lon Nordeen, *Fighters over Israel* (New York: Orion Books, 1990), 64.
23. Jeffrey Ethell and Alfred Price, *Air War South Atlantic* (London: Sidgwick and Jackson, 1983), 215; Cdr."Sharkey" Ward, *Sea Harrier over the Falklands* (London: Orion, 1993).
24. David Hughes, "USAF Uses Sparrows and Sidewinders in Successful Attacks on Iraqi MiG's," *Aviation Week and Space Technology,* 4 February 1991, 68–69; David C. Isby, *Jane's Fighter Combat in the Jet Age* (London: Harper-Collins, 1997), 164. Another interesting observation is that allied forces expended 178 air-to-air missiles to shoot down 45 Iraqi planes, for a 25 percent SSKP (165).

Chapter 17. Building People at China Lake

1. The list of China Lake recruiting techniques is based on a telephone interview with Lester Garman, 27 December 1992.
2. Telephone interview with Bernard Jaeger, 18 February 1990, supplemented by a letter 21 February 1994.
3. Interview with Thomas Amlie, 23 March 1988, and letter 30 October 1994.
4. Interview with William Woodworth, 21 August 1992.
5. Interview with William Woodworth, 21 August 1992. In a later interview, 24 March 1993, Jack Crawford said he thought Woodworth had been a good manager.
6. Interview with Charles P. Smith, 27 March 1987.
7. Interview with Walter LaBerge, 22 September 1990.
8. While some people called McLean "Bill," others did call him "Doctor," especially when he became technical director. This may have been a special honorific, rather than a routine recognition of his Ph.D.
9. Telephone interview with Warren Legler, 5 February 1989. During China Lake's fiftieth anniversary celebrations in November 1993, at a banquet with about two thousand attendees, representatives of four major categories of China Lakers were honored: the researchers, the technicians, the spouses, and the military. The longest-serving technician was honored at the same time, and in much the same way, as LaV McLean.
10. Interview with Frank Knemeyer, 21 February 1995.
11. Telephone interview with Lt. Col. Thomas McElmurry, 4 December 1993.
12. Telephone interview with Earle Mayfield, 20 February 1994.

13. Interview with Lee Sutton, 5 November 1993; interview with Madelyn Fortune, 22 February 1995.
14. Interview with Darryl Baxter, 5 November 1993.
15. Charles P. Smith, letter to the author, 13 June 1996.
16. Telephone interview with Earl Donaldson, 6 February 1994.
17. Interview with Robert G. S. Sewell, 23 February 1993, 53.
18. Interview with Henry Blazek, 5 March 1994.
19. Rene Vallery-Radot, *The Life of Pasteur* (New York: Dover, 1948), 76.   [309]
20. Interview with Robert G. S. Sewell, 23 February 1993, 27, 58.
21. Jacob Rabinow, "Rabinow Laws," undated typescript.
22. Interview with Robert G. S. Sewell, 23 February 1993, 53–54.
23. Frank Knemeyer, personal communication.
24. Interview with Lucien Biberman, 23 July 1991.
25. Telephone interview with Bernard Smith, 17 September 1994; telephone interview with Lillian Regelson, 17 March 1996.
26. Interview with Franklin Knemeyer and Leroy Riggs, 22–23 August 1995.
27. Information in the remaining part of this section is from an interview with Fred Davis, 8 January 1988.
28. Arthur Squires, *The Tender Ship: Government Management of Technology* (Boston: Birkhauser, 1986).
29. Interview with Jack Crawford, Naval Weapons Center Oral History Interview S-171 (China Lake, California), 20 September 1988, 62.
30. Ernest G. Cozzens (then associate head of the NWC Engineering Department), oral history interview with Leroy Doig III, NWC Oral History S-126, 25 June 1981, 40.
31. Interview with Edward E. Benton Jr., by Ron Westrum and Elizabeth Babcock, China Lake Oral History S-193, 26 August 1991.
32. Interview with Frank Knemeyer, 20 February 1995.
33. Howard A. Wilcox, oral history interview with Elizabeth Babcock and Mark Pahuta, NWC Oral History S-196, San Diego, California, 22 October 1991, 27–28. Rear Adm. Walter V. R. Vieweg, NOTS Commander, September 1949–October 1952, was then a captain. Frank Cartwright remembers this event as taking place with Adm. "Deak" Parsons.
34. Ernest G. Cozzens, oral history interview, 25 June 1981, 37–38.
35. Interview with Burrell Hays, 20 February 1995.
36. R. R. Wilson, "My Fight against Team Research," *Daedalus* 99 (1970) 1076–87, quotation on 1081.
37. Sidewinder Group Interview, William F. Wright and Bud Gott, interviewers, China Lake, 14 March 1980, 26.
38. Quotation from Evelyn Glatt, *The Demise of the Ballistics Department* (Berkeley: University of California Institute of Governmental Studies, 1964), 19.

39. Interview with Lillian Regelson, 17 March 1996.
40. Vice Adm. William Moran, comments on draft, 1994.
41. Robert Freed Bales, "Task Roles and Social Roles in Problem-Solving Groups," in *Readings in Social Psychology*, ed. E. E. Maccoby, T. M. Newcomb, and E. L. Hartley (New York: Holt, Rinehart, and Winston, 1958). "Socioemotional leadership" is a small-group concept that I have used in a larger context.
42. Interview with Peter and Polly Nicol, by Elizabeth Babcock, William R. Blanc, Leroy Doig III, and Mark Pahuta, S-225, 17 March 1993, 83.
43. Interview with LaVerne McLean, by William F. Wright, NWC Oral History S-113, 18 March 1980, 20.
44. Interview with LaVerne McLean, 18 March 1980, 21.
45. Interview with Robert G. S. Sewell, 23 February 1993, 59.
46. Interview with Robert G. S. Sewell, 23 February 1993, 59.

Chapter 18. Building Ideas at China Lake
1. Daniel Kevles, *The Physicists: The History of a Scientific Community in Modern America* (Cambridge, Mass.: Harvard University Press, 1987), 324 et seq.; J. Ronald Fox, *Arming America: How the U.S. Buys Weapons* (Boston: Harvard Business School Press, 1974). This tendency for bureaucrats to take over in peacetime was observed centuries ago by the Islamic philosopher Ibn Khaldun, who pointed out that the power of the bureaucrat *(wazir)* was strongest in peacetime. See translation in Ron Westrum and Khalil Samaha, *Complex Organizations: Growth, Struggle, and Change* (Englewood Cliffs, N.J.: Prentice-Hall, 1984), 49. For varying views about weapons procurement, see Jacques Gansler, *Affording Defense* (Cambridge, Mass.: MIT Press, 1991); Asa A. Clark IV, Peter W. Chiarelli, Jeffrey S. McKitrick, and James W. Reed, eds., *The Defense Reform Debate: Issues and Analysis* (Baltimore: Johns Hopkins Press, 1984); Serge Herzog, *Defense Reform and Technology: Tactical Aircraft* (Westport, Conn.: Praeger, 1994).
2. Merton Peck and Frederic Scherer, *The Weapons Acquisition Process: An Economic Analysis* (Boston: Harvard Graduate School of Business Administration, 1962).
3. Gansler, *Affording Defense*, 151; Berkeley Rice, *The C-5A Scandal: An Inside Story of the Military-Industrial Complex* (Boston: Houghton Mifflin, 1971); Wayne Biddle, "The Real Secret of Stealth," *Discover*, February 1986; Tim Weiner, "A Wing and a Prayer," in *Blank Check: The Pentagon's Black Budget* (New York: Warner Books, 1991), 73–107; George C. Wilson and Peter Carlson, "The Ultimate Stealth Plane," *Washington Post National Weekly Edition*, 1–7 January 1996, 4–9. On the legislative problems, see Jacques Gansler, *Affording Defense*, 95–214.

4. For information on whistle-blowers see Andy Pasztor, *When the Pentagon Was for Sale* (New York: Scribners, 1995). Pasztor deals largely with document theft and bid-rigging. In my mind these are minor sins compared with deceptive reporting of weapon performance or the release of flawed systems (such as the M-16 rifle). The smaller sins are crimes against free enterprise; the latter seriously compromise our warfighting ability and in wartime might be treated as sabotage or treason. For examples of such serious breaches, see A. Ernest Fitzgerald, *The High Priests of Waste* (New York: W. W. Norton, 1972) and *The Pentagonists* (Boston: Houghton Mifflin, 1989); James G. Burton, *The Pentagon Wars: Reformers Challenge the Old Guard* (Annapolis, Md.: Naval Institute Press, 1993). These books show that corporate corruption often occurs in partnership with Pentagon bureaus and sympathetic legislators. The virulent reaction to whistle-blowers shown by bureaucrat and industrialist alike suggests how far those involved will go to defend a troubled system. Unjust treatment of whistle-blowers is common. The treatment of Fitzgerald, Burton, and others should shame and anger every American. If this view seems overly rosy about R&D in World War II, one has only to recall the development times for the following classic WWII vehicles: the Jeep (15 months), the Weasel (7 months), and the DUKW (8 months). See James Phinney Baxter, *Scientists against Time* (Cambridge, Mass.: MIT Press, 1968). Compare these to the years it has taken to evolve the Hummer, replacement for the Jeep. [311]

5. Interview with Vice Adm. William J. Moran by Leroy Doig III, NWC Oral History S-187, 1990. It is interesting to compare the situation in defense projects with "positive bootlegging" in U.S. research in physics. See Richard Muller, "Innovation and Scientific Funding," *Science* 209 (22 August 1980). The Lockheed Skunk Works and the Area 51 Test Center near Groom Lake, Nevada, are associated with many of the "black" aviation programs. While the output of the Skunk Works is highly respected, black programs also include such questionable contributions as the notorious A-12 attack plane, on which the taxpayers have spent $3 billion without getting so much as a prototype. See Wilson and Carlson, "Ultimate Stealth Plane."

6. James W. Canan, *The Superwarriors: The Fantastic World of Pentagon Superweapons* (New York: Weybright and Talley, 1975).

7. Gansler, *Affording Defense,* 142–43; Robert Perry, Giles K. Smith, Alvin J. Harman, and Susan Henrichsen, "System Acquisition Strategies," Rand R-733-PR/ARPA, June 1971. The Russians, notably, have a very different view of quantity versus quality, preferring large numbers of machines to spectacular performance of a few units.

8. Such alternatives were discussed during the hearings on Weapons Systems Acquisition Process, before the Committee on Armed Services of the U.S. Senate, 92d Congress, December 1971. A series of studies by the Rand Corporation is valuable in outlining some of these alternatives and their benefits: see Perry et al., "System Acquisition Strategies"; Burton H. Klein, T. K. Glennan Jr., and G. H. Shubert, "The Role of Prototypes in Development," RM-34671-PR, April 1971; G. K. Smith, A. A. Barbour, T. L. McNaugher, M. D. Rich, and W. L. Stanley, "The Use of Prototypes in System Development," R-2345-AF, March 1981. Ralph Katz, in his "Managing Technological Leaps: A Study of DEC's Alpha Design Team," *Research in Organizational Change and Development* 7 (1993), found a skunk works approach key to development of the Alpha RISC chip. But even on small projects, where skunk works have their greatest advantages, there are some dissenters. It is difficult to set aside the judgment of Mariann Jelinek and Claudia Bird Schoonhoven, *Innovation Marathon: Lessons from High Technology Firms* (Oxford: Basil Blackwell, 1990), whose fine study did not find skunk works critical for innovation in electronics companies. See also Henry Petroski, "Boeing 777," *American Scientist* (November–December 1995): 519–22.

9. The formulation suggested here is mostly my own interpretation, but the fifth principle was suggested by former NOTS project manager Harold Metcalf in a letter of 19 November 1993, responding to an earlier draft. Metcalf made a career at China Lake of being a "number two" person to many of the maestros and studied their methods carefully. He was one of only two people that Franklin Knemeyer ever allowed to use his signature (personal communication, Franklin Knemeyer).

10. Interview with Donald Moore, 20 August 1989.

11. Cf. J. W. Getzels, "Problem Finding and Creative Thought," *Questioning Exchange* 2, no. 2 (1988): 95–103.

12. William B. McLean, Naval Weapons Center Oral History Interview by Albert Christman, 5313-S-97, July 1975, 11.

13. Interview with Robert G. S. Sewell, 23 February 1993, 49.

14. This is awfully close to the "hacker ethic" as expounded in Steven Levy, *Hackers: Heroes of the Computer Revolution* (Garden City, N.Y.: Doubleday, 1984). For more on the culture of innovation in high-tech organizations, see Jelinek and Schoonhoven, *Innovation Marathon*.

15. Interview with Ernest G. Cozzens by Leroy Doig III, 25 June 1981, NWC Oral History S-126, 40. Commentators on this book in manuscript noted that weapons safety was given a high priority at NOTS, and that, if Newt Ward bent the rules, there had to be a very good reason for it.

16. Interview with Roderick McClung by Elizabeth Babcock, 22 and 26 March 1991, 62–63. Howard Wilcox commented that similar sentiments pervaded Los Alamos during World War II.

17. Telephone interview with Bernard Jeager, 18 February 1990.

18. Interview with Roderick McClung, 18 February 1995; interview with Roderick McClung by Elizabeth Babcock, NWC Oral History S-188, 1991, 37.

19. Lillian Regelson, letter to the author, 14 May 1996.

20. Telephone interview with Capt. Frank Ault, 7 May 1993. In his well-known study of the background of important weapons systems ("Hind-sight"), Chalmers Sherwin found that a majority of the projects had used local funds, often bootlegged, to build a prototype or carry out a feasibility study (Sherwin, "Multiple Weapon System Advances through Multiple Innovations," in *Climate for Creativity*, ed. Calvin Taylor [New York: Pergamon, 1972], 143).

21. Industrial sociologists refer to such knowledge groups as a "community of practice." See John Seely Brown and Paul Duguid, "Organizational Learning and Communities-of-Practice: Toward a Unified View of Working, Learning, and Innovation," *Organization Science* 2, no. 1 (1991). Many interviewees expressed anger over Freeman's attempts to bring "real management" to China Lake. According to Leroy Doig II, many of the bulky official rules by which procurement was supposed to operate were responses to earlier China Lake transgressions (personal communication from his son, Leroy Doig III).

22. Interview with Jack Crawford, Naval Weapons Center Oral History Interview S-171, China Lake, California, 20 September 1988, 74.

23. Interview with LaVerne McLean, by William F. Wright, NWC Oral History S-113, 18 March 1980, 2; telephone interview with Bernard Jaeger, 18 February 1992.

24. Teresa Amabile used this title for a paper presented at the conference on social psychology of science, Atlanta, Georgia, 1990.

25. Interview with Ken Powers by Elizabeth Babcock, 18 May 1981, NWC Oral History S-123, 1–11. Robert Blaise, who worked in the Pentagon after his years at China Lake, confirmed Powers's observation in regard to such laboratories as the Naval Weapons Laboratory at Dahlgren and the Naval Ordnance Laboratory at White Oak, where socialization after work was unusual (telephone interview with Robert B. Blaise, 22 November 1992).

26. Telephone interview with Bernard L. Smith, 17 September 1994.

27. The rapport between China Lake's hands-on developers and Pentagon "whiz kids" was not good. When one of Herbert York's whiz kids suggested that success of a satellite program could be measured by pounds in orbit divided by dollars, Smith suggested orbiting concrete. This

statement was not well received. (Telephone interview with Bernard
L. Smith, 17 September 1994.)

28. Telephone interview with Bernard L. Smith, 17 September 1994.
29. Interview with Ernest G. Cozzens by LeRoy Doig III, 25 June 1981.
30. Interview with Franklin Knemeyer and Leroy Riggs, 22 August 1995.
31. Information on Rexroth's role was developed through interviews with Burrell Hays, Frank Knemeyer, and Leroy Riggs.
32. Paisely's activities are covered extensively in Andy Pasztor's *When the Pentagon Was for Sale.* Barely mentioned in the book is Tacit Rainbow, an antiradar missile that was one cause of Burrell Hays's forced resignation from his post of technical director at China Lake (Ralph Vartebedian, "Paisely Halted Experts' Attempts to Kill Missile," *Los Angeles Times,* 30 June 1988).
33. Interview with Donald Moore, 20 August 1989.
34. Interview with Donald Moore, 20 August 1989.
35. Telephone interview with Robert R. Sizemore, 22 February 1994.

Chapter 19. The Past and Future Laboratory

1. The suggestion for a "dogfight missile" is recommendation no. 17 under item 4 "Production vs. Design" in Naval Air Systems Command, *Report of the Air-to-Air Missile System Capability Review,* July–November 1968, 1:31. Thrust-vector attitude control is not mentioned in the report, although Ault may have suggested it orally. Ault's assertion that this was the most important recommendation of the report is not apparent in the report itself.
2. Interview with Vice Adm. William J. Moran, 25 October 1994.
3. Interview with Capt. Frank Ault, 23 April 1994; Michael A. Dornheim and David Hughes, "U.S. Intensifies Effort to Meet Missile Threat," *Aviation Week and Space Technology,* 16 October 1995, 36–39.
4. Interview with Robert G. S. Sewell, 23 February 1993.
5. Interview with Richard Schmitt, 24 August 1995.
6. C. Northcote Parkinson, *Parkinson's Law and Other Studies in Administration* (New York: Ballantine, 1964).
7. Interview with LeRoy Riggs, 16 January 1988.
8. This section is based on a manuscript by Leroy Doig III that was given by him to the author, 24 March 1987.
9. Interview with Leroy Riggs by LeRoy Doig III, June 1983, cited in Leroy Doig III MS.
10. Interview with LaVerne McLean, 21 August 1995.
11. Interview with Barney Towle, 23 July 1988.
12. Interview with Capt. Charles Bishop, 16 January 1988.
13. Ann Markusen et al., *The Rise of the Gunbelt: The Military Remapping of America* (New York: Oxford University Press,1991), 225.

14. Eugene Ferguson, *Engineering and the Mind's Eye* (Cambridge, Mass.: MIT Press, 1992); James Fallows, *National Defense* (New York: Random House, 1981).

15. Wilcox's comments made during interview with Donald Moore, 20 August 1989.

16. Robert Frank Futrell, *Ideas, Concepts, Doctrines: Basic Thinking in the United States Air Force, 1907–1960,* vol. 1 (Maxwell Air Force Base, Ala.: Air University Press, 1989), 486–87.

17. Cf. John H. Troll, "Of Sidewinders and Dead Ends," *Reporter,* 13 November 1958, 21–24. In part this willingness to transfer thinking to industry reflected belief in team research. R&D was seen as a commodity that could be purchased by the pound. There has been an effort to minimize the role played by hunch, intuition, and the actions of individuals and small-scale firms—a fatal error.

18. Telephone interview with LeRoy Riggs, 10 January 1995.

19. Telephone interview with Norman Woodall, 25 February 1996.

20. See Charles J. V. Murphy, "Planemakers under Stress," *Fortune,* June 1960, 134–204, and July 1960, 111–244. Of course China Lake did not compete for production contracts. But Richard Schmitt, a design engineer both at NOTS/NWC and later in industry, said his industry experience made him realize that a design team at NWC could mean unemployment to his counterparts at aerospace firms (interview with Richard Schmitt, 24 August 1995).

21. Interview with Jack Crawford, NWC Oral History, 20 September 1988.

22. Interview with Frank Knemeyer, 20 February 1995.

23. Interview with Jack Crawford, NWC Oral History, 20 September 1988, 66; emphasis added.

24. See James G. Burton, *The Pentagon Wars: Reformers Challenge the Old Guard* (Annapolis, Md.: Naval Institute Press, 1993), 217–18. This book provides an important perspective on what kind of honesty the government can expect from its contractors and program managers without watchdogs like China Lake.

25. L. A. "Pat" Hyland, *Call Me Pat* (Virginia Beach: Donning, 1993), 271.

26. Interview with Burrell Hays, 18 February 1995.

27. Conrad R. Hilpert, "A Unique View of Academe," *Bent* (winter 1995): 19–22.

28. Thomas S. Amlie, to Head, Aviation Ordnance Dept., "Trip Report," 354/TSA:drq, Serial 26, 13 May 1960, 3.

29. Bill Gunston, *The Illustrated Encyclopedia of the World's Rockets and Missiles* (London: Salamander Books, 1979), 123; Fred Alpers, Vernon Dell, and James McLane at the Naval Ordnance Laboratory Corona had pioneered the "convert-a-bomb" system that added guidance units and

control surfaces to ordinary iron bombs. While some think these weapons (and their descendants) were as good as Walleye, this is a moot point. NOL Corona became part of China Lake in 1971.

30. Ben Rich, *Skunk Works: A Personal Memoir of My Years at Lockheed* (Boston: Little, Brown, 1994).

31. See Burton, *Pentagon Wars.*

32. James Stevenson, *The Pentagon Paradox: The Development of the F-18 Hornet* (Annapolis, Md.: Naval Institute Press, 1993).

33. This point was made very forcibly to me by Max White, the historian at Point Mugu, during an interview on 17 June 1995.

34. Personal communication from Max White.

Index

ABOUT THE AUTHOR

Ron Westrum is professor of sociology and interdisciplinary technology at Eastern Michigan University. He graduated with honors from Harvard and earned a Ph.D. in sociology from the University of Chicago. He has written two previous books, *Complex Organizations: Growth, Struggle, and Change* (with Khalil Samaha) and *Technologies and Society: The Shaping of People and Things,* as well as numerous articles in the area of science, technology, and society. Professor Westrum's work in aviation and systems safety has received worldwide recognition, and he is a frequent speaker at international conferences. *Sidewinder* is the result of twelve years of research on the highly creative culture of the China Lake Naval Weapons Center.

The Naval Institute Press is the book-publishing arm of the U.S. Naval Institute, a private, nonprofit, membership society for sea service professionals and others who share an interest in naval and maritime affairs. Established in 1873 at the U.S. Naval Academy in Annapolis, Maryland, where its offices remain today, the Naval Institute has members worldwide.

Members of the Naval Institute support the education programs of the society and receive the influential monthly magazine *Proceedings* and discounts on fine nautical prints and on ship and aircraft photos. They also have access to the transcripts of the Institute's Oral History Program and get discounted admission to any of the Institute-sponsored seminars offered around the country.

The Naval Institute also publishes *Naval History* magazine. This colorful bimonthly is filled with entertaining and thought-provoking articles, first-person reminiscences, and dramatic art and photography. Members receive a discount on *Naval History* subscriptions.

The Naval Institute's book-publishing program, begun in 1898 with basic guides to naval practices, has broadened its scope in recent years to include books of more general interest. Now the Naval Institute Press publishes about one hundred titles each year, ranging from how-to books on boating and navigation to battle histories, biographies, ship and aircraft guides, and novels. Institute members receive discounts of 20 to 50 percent on the Press's nearly eight hundred books in print.

Full-time students are eligible for special half-price membership rates. Life memberships are also available.

For a free catalog describing Naval Institute Press books currently available, and for further information about subscribing to *Naval History* magazine or about joining the U.S. Naval Institute, please write to:

Membership Department
U.S. Naval Institute
291 Wood Road
Annapolis, MD 21402-5034
Telephone: (800) 233-8764
Fax: (410) 269-7940
Web address: www.usni.org